A World Synopsis of the Genus *Grimmia*

(Musci, Grimmiaceae)

Jesús Muñoz
Francisco Pando

A WORLD SYNOPSIS OF THE GENUS *GRIMMIA*

(Musci, Grimmiaceae)

Jesús Muñoz
Francisco Pando

83

Missouri Botanical Garden Press

MONOGRAPHS IN SYSTEMATIC BOTANY
FROM THE MISSOURI BOTANICAL GARDEN

Volume 83

ISBN 0-915279-92-4
ISSN 0161-1542

AUTHORS' ADDRESS

Real Jardín Botánico
Plaza de Murillo 2
E-28014 Madrid, Spain.
jmunoz@ma-rjb.csic.es
pando@ma-rjb.csic.es

The cover and title page illustration is
Grimmia trinervis R. S. Williams, by David Foronda (Madrid).

P.O. Box 299
St. Louis, Missouri 63166-0299, U.S.A.

Printed in U.S.A.

She's anything but typical...

She's a powerful force...

She's a natural law...

She used to look good to me, but now I find her

Simply irresistible

Robert Palmer
Simply Irresistible

From the CD "Heavy Nova"

Table of Contents

ABSTRACT

The genus *Grimmia* is considered to comprise 71 species. In this synopsis no taxa at infraspecific rank are recognized. For each accepted species synonyms are listed, mostly based on the study of their type material. 131 names names are lectotypified, and references to previous correct lectotypifications are given where appropiate. Also presented is a key and the distribution of each species, based solely on specimens studied by the authors. Names for which type material could not be studied yet are listed in a different section with the geographical information of the protologue to facilitate the location of type specimens.

Some widely used names are shown to be mere synomyms of others, mostly from different geographical regions. Among those worth mentioning are the following, with the accepted name in parentheses: *G. affinis* Hornsch. (= *G. longirostris* Hook.); *G. apiculata* Hornsch. (= *G. fuscolutea* Hook.); *G. arizonae* Renauld & Cardot (= *G. pilifera* P. Beauv.); *G. brittoniae* R.S. Williams (= *G. funalis* (Hedw.) Schwägr.); *G. curvata* (Brid.) De Sloover and *Dryptodon patens* (Hedw.) Brid. (= *G. ramondii* (Lam. & DC.) Margad.); *G. hermanii* Crum (= *G. muehlenbeckii* Schimp.); *G. moxleyi* R.S. Williams (= *G. orbicularis* Wilson); *G. sessitana* De Not. (= *G. reflexidens* Müll. Hal.); *G. tenerrima* Renauld & Cardot (= *G. reflexidens* Müll. Hal.); *G. tolucensis* Cardot (= *G. elongata* Kaulf.).

ACKNOWLEDGEMENTS

Jesús Muñoz expresses his gratitude to the staff of the Missouri Botanical Garden, and especially his colleagues at the bryological laboratory, for providing a nice working environment during his two-year stay. Most of the work was done in St. Louis during a post-doctoral stay funded by the Spanish Ministry of Education and Culture. He is also grateful to the curators of the herbaria cited in the text for their kind assistance during his visits and for the loans of specimens. The visit to Helsinki (H) was funded by the European Commission HCM Contract nl ERBCHGECT940065 with the Division of Systematic Biology of the University of Helsinki. The Friends of the Farlow and the W.C. Steere Fund supported his visits to Harvard University (FH) and The New York Botanical Garden (NY), respectively. Travels to Jena (JE) and Paris (PC) during 1999 were funded by the Missouri Botanical Garden. He is also indebted to Rosa M. Ros and M. Jesús Cano (Murcia, Spain), for inviting him on several collecting trips that have increased the known geographic range of several species; and to Marite Tellería (Madrid) for her unconditional support. Steve Churchill (St. Louis) and Carlos Aedo (Madrid) deserve special mention: they persuaded Muñoz to publish parts of this work and have supported and helped him in many ways.

We thank Marshall R. Crosby (St. Louis), Félix Muñoz (Madrid), Ryszard Ochyra (Krakow), Marko Sabovljević (Belgrade), and Alan Whittemore (St. Louis, now Washington, D.C.) for help with literature or doubts about nomenclatural puzzles; Si He (St. Louis), Masanobu Higuchi (Tokyo), and Tatyana V. Shulkina (St. Louis) for translations from Chinese, Japanese, and Russian, respectively. We also thank Olga Afonina (St. Petersburg), Jan-Peter Frahm (Bonn), Harald Kürschner (Berlin), and Beata Papp (Budapest) for providing specimens from areas little known to the authors.

INTRODUCTION

Grimmia is the largest genus in the Grimmiaceae. It is considered a difficult genus, and its species are believed to be protean in nature. The combination of a large number of species and the absence of treatments to name extra-European specimens has hindered many from dealing with the genus in any depth.

For several years, *Grimmia* has been the research of one of the authors (JM). What started as a Ph.D. subject (*Grimmia* subg. *Orthogrimmia*, see Muñoz, 1998b), soon evolved into a world revision of the genus. It also soon became evident that the large number of described taxa made further study an impossible task without the aid of an electronic database. The second author (FP) had designed at that time a database to store names and a program to automatically generate nomenclatural lists. A collaboration produced a checklist of the genus. The names and associated information (protologue, distribution, etc.) was stored in the database system "Bibmaster" (www.rjb.csic.es/bibmaste/bibmaste .htm). The nomenclatural synopsis here presented, along with the name index, was generated automatically from the system, which also provides support for data consistency, publication abbreviations, and other tools for data capture and maintenance.

In this synopsis we recognize 71 species in *Grimmia*. In fact, it represents the nomenclatural treatment of an ongoing world revision of the genus. Its structure is similar to the nomenclatural section in the monographs already published (Muñoz 1998a, 1998b, 1999a). We also present a key to the accepted species and the known world distribution for these taxa. In a time when biodiversity is one of the magic words in biology, we think that presenting authenticated data on the number and distribution of species in the large moss family Grimmiaceae will be helpful in related fields, e.g., bryogeography.

Figure 1 shows the number of species recognized in each area according to the system proposed by Wijk et al. (1962). As it is clear from this figure, *Grimmia* has mainly an holarctic distribution, and it is especially diverse at high latitudes in the Northern Hemisphere. The Southern Hemisphere has a relatively low *Grimmia* diversity, although the total number in some areas could lead to a different conclusion. Larger areas in the Southern Hemisphere are poor in species in fact, and the number only increase at high elevations or latitudes.

The history and the different infrageneric classifications proposed for *Grimmia* were reviewed by Deguchi (1978). The main infrageneric classifications to date were also summarized by Muñoz (1998b: table 1). A new, rather eclectic, infrageneric classification, has been published by Nyholm (1998) in her last treatment of the Fennoscandian species. As already stated (Muñoz, 1998b: 374), the infrageneric classification of *Grimmia* has yet to be resolved, and therefore we deliberately do not embrace any of those proposed systems.

MATERIALS AND METHODS

We compiled the names of *Grimmia*, *Dryptodon*, and other related genera accepted as pertaining to *Grimmia* in *Index Muscorum* (Wijk et al., 1962, 1969), but not those that *Index Muscorum* gives as accepted in a different genus (e.g., *Grimmia flexuosa* Griff., for which the accepted name in *Index Muscorum* is *Garkea comosa* (Dozy & Molk.) Wijk & Margad., was not included in the database). Afterward, all *Grimmia* names found in Crosby et al. (1992), Crosby and Magill (1994), and Crosby and Magill (1997), or published thereafter were included in the nomenclatural database. Hence, the database does not list all names published in the genus, and *Index Muscorum* has to be consulted for those names not appearing here. However, a few taxa originally excluded from *Grimmia* in *Index Muscorum* have been resurrected and included here. The database contains some 1370 names, which correspond to 792 taxa. The protologue of each name has been checked, involving the review of more than 500 different papers. Type or original material of 590 taxa representing more than 74% of the total have been studied so far. Among the taxa for which types have not been studied, there are 65 (8.2%) described at the species level and 139 (17.55%) described at the variety or form level. From a preliminary survey of these taxa not yet studied, none threatens to be older than any accepted name in this work, which assures nomenclatural stability.

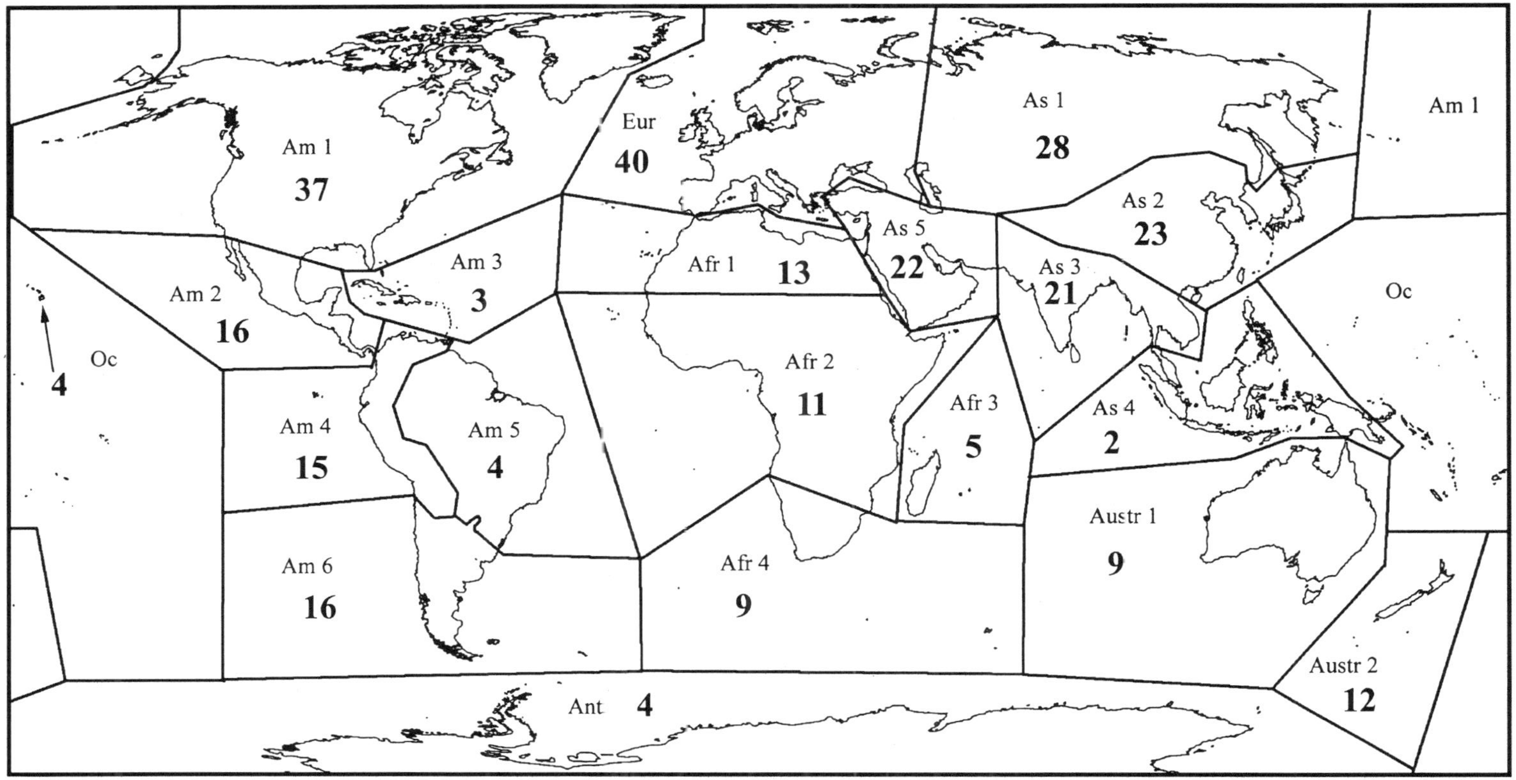

Figure 1. Number of *Grimmis* spec:es in the geographical areas proposed by Wijk et al. (1962).

The work is divided into three sections:

1. **Synopsis of the genus *Grimmia***. This includes the names for which we have seen the type(s) or, in a few cases, accurate and informative illustrations of types (e.g. Deguchi, 1978; Geissler & Maier, 1995; Maier, 1998). In this section accepted names, in **bold**, are listed alphabetically, and synonyms, in *italic*, are listed under the accepted name in chronological order of publication. Distribution is indicated by citing a single collection from each country.

2. **Taxa excluded from *Grimmia***. Taxa belonging to other genera are included here. The structure of this section is the same as that of the previous section. Some names belonging in *Racomitrium* and *Schistidium* have been transferred to these genera by other authors, and we follow their views regarding generic placement, but we have no opinion on the correct species name, because these combinations are not made in the context of a taxonomic revision. Consequently, those names are listed alphabetically under ***Racomitrium* spp**. and ***Schistidium* spp**. Both genera are in urgent need of taxonomic revision outside Europe, and any decision now seems premature.

3. **Names with type material not seen**. In this section we list alphabetically those names for which we have not seen the type(s). All names have been at some moment included in *Grimmia*, or were described as infraspecific entities of taxa now considered to belong in *Grimmia*. Some, if not many, of the names here included have been considered synonymous with other taxa by previous authors (see Wijk et al., 1962, 1969). We do not present here any information on previous synonymizations, since many of those have proved to be wrong. This does not mean that we consider any of them "good" taxa. Surely these taxa will be synonyms, but of different taxa from which they will use to be considered conspecific. Although apparently no name in this section threatens any of the accepted names in the first section, we continue to reduce the size of this section.

Every entry follows the same scheme: name, author(s), place of publication, homotypic synonyms ordered chronologically, and type locality according to the protologue. The type locality according to the protologue ("PROTOLOGUE" in the text) is the exact transcription of the locality in the protologue. It has been taken directly and has not been edited. If further information has been added, it is placed between square brackets.

We have treated all *nomina nuda* (nom. nud.), and placed them in the species lists. These names are invalid and accordingly have no types. Therefore, we have included the geographical information on these names under the header "ORIGINAL MATERIAL," and the information on the specimen labels, under "ORIGINAL SPECIMENS." In a few instances (e.g., *G. biplicata*, *G. meridensis*, and *G. paramophila*), no original specimen(s) was seen, and our judgment is based on illustrations or comments on the original publication.

For names in the first two sections, "TYPE" is the transcription of the type (holotype or lectotype) label, and herbarium(a) where the type(s) is housed. The protologue and label information are included for verification of our typifications. Bibliographic references to lectotypifications by other authors have also been indicated where appropriate. We have selected new lectotypes for all names not typified previously.

Distributions are given only for accepted names. They are based solely on studied specimens. Some 10,000 specimens have been studied so far, and one representative specimen is listed together with the herbarium where this voucher is located. We have used ISO-3166 standard for country codes.

The only exception is Russia (ISO code RUS, not used here), which has been divided into three areas according to Ignatov and Afonina (1992): European Russia (SUN); Siberia (SIB); and Far East (SFE).

Areas in the Near East, Siberia, and Africa are underrepresented, and any specimen from those areas is welcomed by the first author.

GRIMMIA HEDW.

Sp. Musc. Frond.: 75. 1801.

Type species: *Grimmia plagiopoda* Hedw. (lectotype, designated by Mårtensson, 1956: 156).

Dryptodon Brid., Bryol. Univ.: 191. 1826.

Type species: *Dryptodon patens* (Hedw.) Brid. [≡ *Bryum patens* Hedw.] (lectotype, designated by Ochyra, Sérgio & Schumacker, 1988: 246).

Guembelia Hampe, Bot. Zeitung (Berlin) 4: 124. 1846.

Type species: *Guembelia elliptica* (Funck) Hampe [≡ *Grimmia elliptica* Funck] (lectotype, designated by Pfeiffer, 1874: 1511).

Schistidium Brid., Muscol. Recent. Suppl. 4: 20. 1818 ["1819"]. *Anodon* Rabenh., Deutschl. Krypt.-Fl. 2(3): 153. 1848, nom. illeg.

Type species: *Anodon pulvinatus* (Hedw.) Rabenh. [≡ *Gymnostomum pulvinatum* Hedw.] (designated by Pfeiffer, 1873: 199).

NOTE: Ochyra (1989a) showed that *Schistidium pulvinatum* (Hedw.) Brid., type of both *Anodon* and *Schistidium* Brid., is in fact *Grimmia anodon*, which converts those genera to taxonomic synonyms of *Grimmia*. As a consequence, he proposed for conservation the name *Schistidium* Bruch & Schimp. with a new type (Ochyra, 1989b).

Hydrogrimmia (I. Hagen) Loeske, Stud. Morph. Syst. Laubm.: 108. 1910. *Grimmia* subgen. *Hydrogrimmia* I. Hagen, Kongel. Norske Vidensk. Selsk. Skr. (Trondheim) 1909(5): 35. 1909. *Grimmia* subgen. *Hygrogrimmia* Vilh., Věstn. Král. Ceské Společn. Nauk, Tř. Mat.-Přír. 1924: 6. 1924, nom. inval., orthogr. err. *Grimmia* subsect. *Hydrogrimmia* (I. Hagen) G.N. Jones, in Grout, Moss Fl. N. Amer. 2: 21. 1933. *Grimmia* sect. *Hydrogrimmia* (I. Hagen) Nyholm, Ill. Fl. Nord. Mosses 4: 334. 1998.

Type species: *Hydrogrimmia mollis* (Bruch & Schimp.) Loeske (holotype).

Trigonodictyon Dixon & P. de la Varde, Ann. Cryptog. Exot. 1: 40. 1928.

Type species: *Trigonodictyon indicum* Dixon & P. de la Varde (holotype).

Usmania Laz., Dopov. Akad. Nauk Ukrajins'k. RSR 11: 1039. 1970.

Type species: *Usmania campylopoda* Laz. (holotype)

Cladautoicous, *gonioautoicous*, or *dioicous*. *Plants* in dense cushions or compact to loose tufts, glaucous, green, greenish yellow or olive-green. *Stems* erect or ascending, with or without central strand. *Leaves* erect, appressed or flexuous, occasionally with homomallous tips when dry, erect to spreading when moist, linear, ovate, lanceolate, ligulate, oblong to triangular, obtuse to acuminate, concave, canaliculate or keeled, plane or plicate; *margins* entire, plane, recurved or incurved; *costa* single, percurrent, terete, semiterete, semi-elliptic or almost indistinct in cross section; *lamina* 1–4-stratose in the distal half; *distal cells* isodiametric to rectangular or oblate, bulging or plane, smooth or papillose, with straight or sinuous walls; *proximal cells* quadrate to rectangular or oblate, the walls straight or sinuous, uniformly thickened or the transverse walls thicker than the longitudinal walls; with or without *hyaline hair-points*. *Perichaetial leaves* convolute and larger or similar in shape but slightly larger than vegetative leaves. *Androecia* axillary or terminal. *Setae* straight or curved, longer or shorter than capsules. *Capsules* immersed, emergent or exserted, subglobose, ovoid, ellipsoid or fusiform, symmetric or asymmetric and ventricose at the base, smooth or ribbed, stramineous or castaneous, with stomata at the urn base or lacking stomata; *exothecial cells* isodiametric or rectangular, thin- or thick-walled; *annulus* simple and persistent, compound and revoluble, or compound and tardily deciduous; *peristome teeth* 16, triangular, entire, perforate in the distal half or cribrose throughout and irregularly cleft at apex, orange to brown; *opercula* conic to long-rostrate; *calyptrae* cucullate or mitrate, smooth, covering at least to the capsule mouth; *columella* persistent; *spores* smooth to coarsely granulose.

Grimmia is a genus of epilithic mosses (except *G. pitardii*, which grows on ground) closely related to *Schistidium* Bruch & Schimp. Indeed, American authors tend to unite both genera (e.g., Crum, 1994, who also joined *Jaffueliobryum* Thér. in his encompassing concept of *Grimmia*). As I conceive the genus, it includes the taxa of Grimmiaceae subfam. Grimmioideae in the sense of Churchill (1981), except *Schistidium*. It could be argued that this approach renders *Grimmia* paraphyletic, but our preliminary results of morphological cladistic analysis support that *Schistidium* is sister to the rest of the subfamily. This phylogenetic analysis of *Grimmia*, including a study of infrageneric relationships, will be presented elsewhere.

Schistidium differs from *Grimmia* basically in two characters: systylius capsules and small calyptrae not reaching the capsule mouth. A character widely used, perichaetial leaf size, must be considered uninformative, because those leaves are also greatly enlarged in most species of *Grimmia*.

KEY TO THE SPECIES OF *GRIMMIA*

Notes for using the key (see Muñoz, 1999a: 121):

Leaf length is *without* hair-points.

Cross sections are essential to correctly identify specimens of Grimmiaceae.

Perichaetial leaf area is considered the area of the rectangle defined by the length and width of the perichaetial leaves, again *without* hair-points.

Capsules may be immersed and are sometimes overlooked. Using the dissecting microscope look carefully for inflated distal stem areas, where leaves have longer hair-points.

1a. Leaves strongly plicate on both sides of costa.
- 2a. Seta sigmoid to coiled; capsule asymmetric and ventricose at base, seta eccentrically placed at base, gymnostomous; calyptrae mitrate, not exceeding further down than capsule mouth, not plicate .... 68. *G. trinervis*
- 2b. Seta straight; capsule symmetric with seta centrally placed at base, peristome present; calyptrae cucullate or, if mitrate, campanulate, covering most of the capsule and plicate.
  - 3a. Capsule exserted, calyptra cucullate, not plicate; laminal cells usually papillose .... 11. *G. caespiticia*
  - 3b. Capsule immersed, calyptrae mitrate, campanulate, plicate; laminal cells not papillose .... [*Coscinodon*]

1b. Leaves not or only weakly plicate.
- 4a. Capsule asymmetric and ventricose at base; seta eccentrically attached to the capsule, sigmoid, shorter than the capsule.
  - 5a. Proximal marginal cells of leaves hyaline, inflated, with all walls similar, thin .... 25. *G. incrassicapsulis*
  - 5b. Proximal marginal cells of leaves not hyaline or inflated, with transverse walls thicker than longitudinal walls.
    - 6a. Leaves obovate, wider at apex; calyptrae cucullate .... 13. *G. crinita*
    - 6b. Leaves subulate to lanceolate, wider at proximal half; calyptrae mitrate.

7a. Costa elliptical in cross section, scarcely prominent on dorsal surface, occasionally of the same thickness as lamina, scarcely differentiated from lamina.
8a. Autoicous, male buds small and hidden just below perichaetia............... 2. *G. americana*
8b. Dioicous, male buds terminal..55. *G. poecilostoma*
7b. Costa semicircular in cross section, prominent on dorsal surface, differentiated from lamina.
9a. Peristome lacking .... 3. *G. anodon*
9b. Peristome present.
10a. Leaves muticous .... 53. *G. pitardii*
10b. Leaves with hair-point... 54. *G. plagiopodia*
4b. Capsule symmetric; seta centrally attached to the capsule, straight or curved but not sigmoid, longer than the capsule, or sterile specimens [NOTE: Not all sterile collections can be identified with confidence].
11a. Costa cross section rounded.. 62. *G. teretinervis*
11b. Costa cross section not rounded.
12a. Proximal marginal cells of leaves oblate; laminal cells not papillose .... 30. *G. laevigata*
12b. Proximal marginal cells of leaves quadrate to rectangular or, if oblate, laminal cells strongly papillose.
13a. Leaves soft, broadly ovate.. 39. *G. mollis*
13b. Leaves rigid, linear, ovate or obovate.
14a. Leaf cells distinctly papillose.
15a. Costa elliptical in cross section, scarcely prominent on dorsal surface, occasionally of the same thickness as lamina, scarcely differentiated from lamina ....36. *G. mammosa*
15b. Costa semiterete to triangular in cross section, prominent on dorsal surface, differentiated from lamina.
16a. Leaves longer than 2 mm; plants robust; stems ascending, to 8 cm long; seta curved .... 17. *G. elatior*
16b. Leaves shorter than 1.8 mm; plants small; stems erect, to 1.5 cm long; seta straight... .... 44. *G. nivalis*
14b. Leaf cells not papillose, pseudopapillose or not.
17a. Proximal marginal cells of leaves with all walls similar, thin.
18a. Margins plane throughout or incurved above, no trace of recurvation.
19a. Leaves crisped when dry...65. *G. tortuosa*
19b. Leaves erect and appressed to weakly flexuous when dry, but never crisped.
20a. Ventral layer of costa 4–6 cells wide .... 71. *G. wilsonii*
20b. Ventral layer of costa 2 cells wide.
21a. Seta straight, to 1 mm; capsule immersed; exothecial cells thick-walled............. .... 67. *G. triformis*
21b. Seta straight or curved, longer than 1 mm; capsule emergent or exserted; exothecial cells thin-walled.
22a. Seta curved; stem leaf hyaline hair-point to 2 mm, those of the perichaetial leaves strongly flexuous and twisted... 6. *G. arenaria*
22b. Seta straight; stem leaf hyaline hair-point to 1 mm, those of the perichaetial leaves weakly flexuous and not twisted .... 16. *G. donniana*
18b. At least one margin recurved for part of its length.
23a. Leaves crisped when dry, narrowly linear-lanceolate from ovate base.
24a. One margin narrowly recurved for the proximal 1/3, the other flat or only recurved for a very short distance from base ....26. *G. incurva*
24b. Both margins recurved to revolute more than half their length... 43. *G. nepalensis*

23b. Leaves straight to flexuous but not crisped, lanceolate to ovate-lanceolate, seldom subulate.
25a. Leaves with an ovate base extended into a subulate distal part mostly filled by the costa; costa excurrent in a strong green mucro .... 26. *G. incurva*
25b. Leaves lanceolate to ovate-lanceolate; lamina evident in distal part of leaves; costa not excurrent as a green mucro.
26a. Capsules cylindrical, 1.5–2.5 mm long .... 35. *G. macrotheca*
26b. Capsules ovoid, ca. 1 mm long.
27a. Longest hair-point longer than 0.3 mm long, most of the leaves piliferous; plants autoicous, male buds small and lateral, hidden just below perichaetia .... 20. *G. fuscolutea*
27b. Longest hair-point to 0.3 mm long, most of the leaves muticous; dioicous, male buds terminal.
28a. Lamina blades in distal half forming <30° angle in cross section, parallel to each other at their junction with the costa, furrow very narrow; seta straight .... 18. *G. elongata*
28b. Lamina blades in distal half forming >45° angle in cross section, not parallel to each other at their junction with the costa, furrow more open; seta curved........ .... 60. *G. redunca*
17b. Proximal marginal cells of leaves with transverse walls thicker than longitudinal walls.
29a. Leaves strongly crisped when dry.. 64. *G. torquata*
29b. Leaves straight, flexuous or curled, but not crisped.
30a. Gemmae present at leaf apices.
31a. Gemmae yellow to yellow-green, brownish when old; ridge-like thickenings over the cell lumen, which appear as papillae in cross section; dorsal side of costa rounded in cross section, without wings or projections .... 4. *G. anomala*
31b. Gemmae reddish; laminal cells without ridge-like thickenings over the lumen; costa usually with wings or projections at dorsal side .... 23. *G. hartmanii*
30b. Gemmae at leaf apices lacking.
32a. Leaves all constantly muticous.
33a. Costa winged on dorsal side .... 59. *G. ramondii*
33b. Costa without projections on dorsal side.
34a. Margins plane, no trace of recurvation.
35a. Leaves homomallous; apex subulate..................................... 21. *G. hamulosa*
35b. Leaves straight; apex obtuse to acute.
36a. Plants brownish; distal part of lamina 2-stratose.............. 46. *G. obtusifolia*
36b. Plants green to black; distal part of lamina 2–5-stratose...... 70. *G. unicolor*
34b. Margins recurved partially to fully on one or both sides.
37a. Lamina 1-stratose, occasionally 1–2 marginal rows 2-stratose.
38a. Leaves lanceolate; apex acuminate; only known sterile....... 22. *G. handelii*
38b. Leaves oblong; apex acute; capsule with a distinct apophysis.................... .... 27. *G. indica*
37b. Lamina 2-stratose for more than the 1–2 marginal rows.
39a. Dry plants with leaves flexuous and crisped; 1 to 2 basal marginal leaf cell rows inflated, usually reddish, forming 2-stratose alar areas; perichaetial leaves virtually identical to vegetative leaves (1.5–3.2 mm vs. 2.3–3.2 mm long); setae 3–10 mm; capsules completely exserted.. 7. *G. atrata*

39b. Dry plants with leaves erect, only flexuous at apical part; basal marginal leaf cell rows not inflated, not colored, similar to adjacent inner rows, 1-stratose; perichaetial leaves longer than vegetative leaves (4.0–5.5 mm vs. 2–3 mm long); setae 1.5–2.1 mm; capsules immersed .... 47. *G. ochyriana*

32b. Leaves, at least perichaetial, with a trace of hyaline hair-point.

40a. Costa elliptical in cross section, scarcely prominent on dorsal surface, occasionally of the same thickness as lamina, scarcely differentiated from lamina.

41a. Proximal juxtacostal cells long rectangular (4–8:1); perichaetial leaves of mature sporophytes green and firm, neither hyaline nor filmy; setae 3–7 mm long; capsule exserted.

42a. Seta curved; basal juxtacostal cell walls not nodulose............... 48. *G. olneyi*

42b. Seta straight; basal juxtacostal cell walls nodulose ....50. *G. ovalis*

41b. Proximal juxtacostal cells shorter (1–4:1); perichaetial leaves of mature sporophytes almost entirely hyaline and filmy (look carefully around capsule, inner perichaetial leaves are almost invisible except under high contrast in the compound microscope); seta to 1 mm; capsule immersed. [NOTE: sterile specimens of these 4 taxa are gametophytically identical and impossible to name with confidence; separation based on ecological preferences is unrealistic].

43a. Capsule asymmetric and ventricose at base; seta sigmoid, eccentrically attached to the capsule.

44a. Plants autoicous, male buds small and hidden just below perichaetia .... 2. *G. americana*

44b. Plants dioicous, male buds conspicuous, terminal on stem......................... ....55. *G. poecilostoma*

43b. Capsule symmetric; seta straight, centrally attached to the capsule.

45a. Plants autoicous, male buds small and hidden just below perichaetia ....28. *G. involucrata*

45b. Plants dioicous, male buds conspicuous, terminal on stem......................... .... 63. *G. tergestina*

40b. Costa semicircular to reniform in cross section, prominent on dorsal surface, differentiated from lamina.

46a. Costa winged, furrowed or irregularly triangular in shape on dorsal side.

47a. Ventral layer of costa 2 to 6 cells wide.

48a. Leaves muticous... 59. *G. ramondii*

48b. Leaves with hair-point .... 23. *G. hartmanii*

47b. Ventral layer of costa 2 cells wide.

49a. Plants to 8 cm long; costa cross section triangular to almost circular, 3–4 layers thick from dorsal to ventral side at midleaf; cells usually papillose, bulging and irregular; gemmae lacking .... 17. *G. elatior*

49b. Plants to 2 cm long; costa cross section flattened dorsally to semiterete, 2(3) layers thick from dorsal to ventral side at midleaf; cells never papillose, bulging or irregular; gemmae common, arising from short stalks in leaf axils.... .... 41. *G. muehlenbeckii*

46b. Costa without wings or projections on dorsal side.

50a. Margins plane throughout or incurved above, no trace of recurvation.

51a. Laminar cells bulging in cross section [NOTE: Sterile or poor specimens of the following three taxa are imposible to name].

52a. Seta less than 1 mm; capsule immersed .... 8. *G. australis*

52b. Seta more than 1.5 mm; capsule exserted.
53a. Stomata lacking at the urn base; capsule usually fusiform, castaneous, concolorous with peristome teeth; exothecial cells isodiametric, thick-walled .... 1. *G. alpestris*
53b. Stomata present at the urn base; capsule ovoid, stramineous, different in color from the orange peristome teeth; exothecial cells irregularly rectangular, thin-walled...61. *G. reflexidens*
51b. Laminar cells not bulging in cross section.
54a. Blades of distal laminal half forming <30° angle in cross section, parallel to each other at their junction with the costa, furrow very narrow.................... ....61. *G. reflexidens*
54b. Blades of distal laminal half forming >45° angle in cross section, not parallel to each other at their junction with the costa, furrow more open.
55a. Distal half of lamina unistratose, only 1 or 2 marginal rows 2-stratose.
56a. Hyaline hair-point terete; seta more than 1 mm, curved; capsule exserted, with peristome.. 24. *G. humilis*
56b. Hyaline hair-point flat at base; seta less than 1 mm, straight; capsule immersed, gymnostomous..56. *G. pseudoanodon*
55b. Distal half of lamina 2–3(4)-stratose.
57a. Lamina concave to U-shaped in cross section in the proximal part of leaf; perichaetial leaves 5–15 times larger in area than other leaves.......... ....29. *G. kidderi*
57b. Lamina V-shaped in cross section in the proximal part of leaf; perichaetial leaves 2–4 times larger in area than the other leaves [NOTE: Sterile or poor specimens of the following four taxa are imposible to name].
58a. Stomata present at urn base.................... 34. *G. macroperichaetialis*
58b. Stomata lacking at urn base.
59a. Leaves ± linear, distal 2/3 subulate; seta to 1 mm; capsule immersed; peristome teeth rudimentary..................... 37. *G. mariniana*
59b. Leaves lanceolate, distal 2/3 acuminate, but not subulate; seta more than 2 mm; capsule emergent to exserted; peristome teeth well developed.
60a. Opercula rostrate; seta 2–4 mm long; peristome teeth 50–90 μm wide at mouth, irregularly splitting above and ± cribrose; proximal juxtacostal leaf cells mostly longly rectangular, to 4.5:1; plants dioicous... 40. *G. montana*
60b. Opercula obtuse to mammillate; seta to 2 mm; peristome teeth 40–50 μm wide at mouth, entire or slightly cribrose at apex; proximal juxtacostal leaf cells isodiametric to shortly rectangular, to 3:1; plants autoicous ....69. *G. ungeri*
50b. Margins recurved partially to fully on one or both sides.
61a. Ventral layer of costa 3–6 cells wide.
62a. Laminal cells bulging; costa semicircular in cross section.. 45. *G. nutans*
62b. Laminal cells not bulging; costa reniform in cross section.
63a. Proximal juxtacostal cells of leaves with uniformily thickened walls.... ....32. *G. lisae*
63b. Proximal juxtacostal cells of leaves with nodulose walls.

64a. Perigonia just below perichaetia; hair-point of apical leaves mostly spreading .... 15. *G. decipiens*
64b. Perigonia terminal; hair-point of apical leaves erect.
65a. Leaves subulate, distal half V-shaped in cross section; costa almost filling the apex .... 5. *G. arcuatifolia*
65b. Leaves acuminate but not subulate, distal half U-shaped in cross section; costa not filling the apex.
66a. Both margins recurved to revolute more than half their length; hyaline hair-point flat; seta curved .... 31. *G. leibergii*
66b. One margin recurved for the proximal 1/2–2/3, the other flat or only recurved for a very short distance from base; hyaline hair-point terete; seta straight .... 33. *G. longirostris*
61b. Ventral layer of costa 2 cells wide.
67a. Laminal cells bulging.
68a. Plants reddish, dioicous .... 51. *G. percarinata*
68b. Plants not reddish, autoicous.
69a. Leaves longer than 2 mm; seta curved .... 45. *G. nutans*
69b. Leaves shorter than 1.8 mm; seta straight .... 61. *G. reflexidens*
67b. Laminal cells not bulging.
70a. Leaves oblong to elliptic, widest near middle or sides parallel, of equal width ± from base to little above middle.
71a. Hyaline hair-point sharply dentate, to 0.4 mm long; plants dioicous .... 57. *G. pulla*
71b. Hyaline hair-point smooth or weakly denticulate, longer than 0.5 mm; plants autoicous.
72a. Peristome teeth vestigial to absent .... 14. *G. curviseta*
72b. Peristome teeth well developed.
73a. Seta to 1 mm long, straight; capsule immersed .... 12. *G. capillata*
73b. Seta longer than 2 mm, curved; capsule exserted
74a. Perigonia just below perichaetia; calyptra mitrate; peristome teeth ± entire .... 58. *G. pulvinata*
74b. Perigonia terminal; calyptra cucullate; peristome teeth cribrose and irregularly divided distally .... 49. *G. orbicularis*
70b. Leaves ovate to lanceolate, widest near base and conspicuously tapering upward.
75a. Dry plants with leaves spirally imbricate along the stem; usually with thread-like shoots with strongly differentiated, muticous and carinate leaves .... 19. *G. funalis*
75b. Dry plants with leaves not spirally twisted along the stem; thread-like shoots lacking.
76a. Lamina 2–3(4)-stratose for most of its upper 2/3 77
77a. Leaves 2–3 mm, ovate, broad proximally, ± suddenly narrowed into an upper subulate part, forming shoulders .... 52. *G. pilifera*
77b. Leaves shorter than 1.9 mm, leaf base not forming shoulders.
78a. Blades of distal laminal half forming <30° angle in cross section, parallel to each other at their junction with the costa, furrow very narrow.

79a. Leaves lanceolate to narrowly lancolate; plants commonly reddish; seta curved...42. *G. navicularis*
79b. Leaves ovate-lanceolate to lanceolate; plants never reddish; seta straight ....61. *G. reflexidens*
78b. Blades of distal laminal half forming >50º angle in cross section, not parallel to each other at their junction with the costa, furrow open.
80a. Leaves straight; setae straight; capsule immersed; plants autoicous ....38. *G. molesta*
80b. Leaves flexuous; setae curved; capsule exserted; plants dioicous ....10. *G. bicolor*
76b. Lamina mostly unistratose in its upper 2/3, with bistratose margins and/or occasional and small 2-stratose patches.
81a. Leaves ovate.. 24. *G. humilis*
81b. Leaves narrowly lanceolate to lanceolate.
82a. Blades of distal laminal half forming <30º angle in cross section, parallel to each other at their junction with the costa, furrow very narrow.. 42. *G. navicularis*
82b. Blades of distal laminal half forming >45º angle in cross section, not parallel to each other at their junction with the costa, furrow more open.
83a. Gemmae sessile on lamina, seldom on the costa, no stalk remaining after gemmae liberation, which causes leaf destruction; lamina bistratose at margins and in streaks; plants less than 3 cm; spores minutely granulose, apparently smooth under light microscope...66. *G. trichophylla*
83b. Gemmae on stalks at dorsal base of costa, the stalk 1(2) cells long remaining attached to the costa after gemmae liberation, which does not affect the leaf; lamina unistratose except sometimes 1 or 2 marginal rows 2-stratose; plants to 8 cm; spores coarsely and distinctly granulose under light microscope....... 9. *G. austrofunalis*

COUNTRY CODES (ISO 3166)

The country codes are listed alphabetically according to the standard ISO-3166. The number of recognized taxa of *Grimmia* known from each country follows between brackets.

AFG Afghanistan (3)
AND Andorra (8)
ARG Argentina (14)
ARM Armenia (14)
ATA Antarctica (4)
ATF French Southern Territories [Croizet Island and Kerguelen Island] (3)
AUS Australia (9)
AUT Austria (27)
AZE Azerbaijan (4)
BEL Belgium (17)
BGR Bulgaria (9)
BIH Bosnia and Herzegovina (1)
BOL Bolivia (14)
BRA Brazil (4)
BTN Bhutan (1)
CAN Canada (27)
CHE Switzerland (30)
CHL Chile (10)
CHN China (16)
CMR Cameroon (2)
COG Congo (1)
COL Colombia (5)
CPV Cape Verde (2)
CRI Costa Rica (2)
CYP Cyprus (6)
CZE Czech Republic (17)
DEU Germany (28)
DNK Denmark (5)
DOM Dominican Republic (2)
DZA Algeria (10)
ECU Ecuador (6)
EGY Egypt (2)
ESP Spain (31)
ETH Ethiopia (4)
FIN Finland (12)
FLK Falkland Islands (1)
FRA France (33)
FRO Faroe Islands (1)
GBR United Kingdom (21)
GEO Georgia (22)
GRC Greece (9)
GRL Greenland (11)
GTM Guatemala (5)
HND Honduras (1)
HUN Hungary (15)
IDN Indonesia (1)
IND India (18)
IRL Ireland (8)
IRN Iran (7)
IRQ Iraq (4)
ISL Iceland (11)
ISR Israel (5)
ITA Italy (31)
JOR Jordan (3)
JPN Japan (10)
KAZ Kazakstan (9)
KEN Kenya (6)
KGZ Kyrgyzstan (6)
KOR South Korea (1)
LBN Lebanon (2)
LBY Libya (1)
LKA Sri Lanka (2)
LSO Lesotho (7)
LVA Latvia (1)
MAR Morocco (10)
MDG Madagascar (2)
MEX Mexico (16)
MKD Macedonia, Former Yugoslav Republic (6)
MNG Mongolia (8)
MWI Malawi (2)
MYS Malaysia (1)
NOR Norway (22)
NPL Nepal (4)
NZL New Zealand (12)
OMN Oman (1)
PAK Pakistan (9)
PAN Panama (1)
PER Peru (10)
PHL Philippines (1)
PNG Papua New Guinea (1)
POL Poland (20)
PRT Portugal (14)
REU Reunion (4)
ROM Romania (11)
SAU Saudi Arabia (6)
SDN Sudan (1)
SFE Russia, Far East (14)
SGS South Georgia (3)
SHN Tristan da Cunha, Ascension, and St. Helena Islands (1)
SIB Russia, Siberia (23)
SJM Svalbard and Jan Mayen Islands (1)
SUN Russia, European (20)
SVK Slovakia (20)
SVN Slovenia (1)
SWE Sweden (22)
SYR Syria (5)
TKM Turkmenistan (8)
TUN Tunisia (7)
TUR Turkey (19)
TWN Taiwan (5)
TZA Tanzania (6)
TZK Tajikistan (5)
UGA Uganda (8)
UKR Ukraine (11)
URY Uruguay (2)
USA U.S.A. (38)
UZB Uzbekistan (6)
VEN Venezuela (6)
VIR Virgin Islands, U.S.A. (1)
YEM Yemen (1)
YUG Yugoslavia (9)
ZAF South Africa (8)
ZAR Zaire [Democratic Republic of Congo] (2)

Alphabetical by country.

Afghanistan	AFG
Algeria	DZA
Andorra	AND
Antarctica	ATA
Argentina	ARG
Armenia	ARM
Australia	AUS
Austria	AUT
Azerbaijan	AZE
Belgium	BEL
Bhutan	BTN
Bolivia	BOL
Bosnia and Herzegovina	BIH
Brazil	BRA
Bulgaria	BGR
Cameroon	CMR
Canada	CAN
Cape Verde	CPV
Chile	CHL
China	CHN
Colombia	COL
Congo	COG
Costa Rica	CRI
Croizet Island	ATF
Cyprus	CYP
Czech Republic	CZE
Denmark	DNK
Dominican Republic	DOM
Ecuador	ECU
Egypt	EGY
Ethiopia	ETH
Falkland Islands	FLK
Faroe Islands	FRO
Finland	FIN
France	FRA
French Southern Territories [Croizet Island and Kerguelen Island]	ATF
Georgia	GEO
Germany	DEU
Greece	GRC
Greenland	GRL
Guatemala	GTM
Honduras	HND
Hungary	HUN
Iceland	ISL
India	IND
Indonesia	IDN
Iran	IRN
Iraq	IRQ
Ireland	IRL
Israel	ISR
Italy	ITA
Japan	JPN
Jordan	JOR
Kazakstan	KAZ
Kenya	KEN
Kerguelen Island	ATF
Kyrgyzstan	KGZ
Latvia	LVA
Lebanon	LBN
Lesotho	LSO
Libya	LBY
Macedonia, Former Yugoslav Republic	MKD
Madagascar	MDG
Malawi	MWI
Malaysia	MYS
Mexico	MEX
Mongolia	MNG
Morocco	MAR
Nepal	NPL
New Zealand	NZL
Norway	NOR
Oman	OMN
Pakistan	PAK
Panama	PAN
Papua New Guinea	PNG
Peru	PER
Philippines	PHL
Poland	POL
Portugal	PRT
Reunion	REU
Romania	ROM
Russia (European)	SUN
Russia (Far East)	SFE
Russia (Siberia)	SIB
Saudi Arabia	SAU
Slovakia	SVK
Slovenia	SVN
South Africa	ZAF
South Georgia	SGS
South Korea	KOR
Spain	ESP
Sri Lanka	LKA
Sudan	SDN
Svalbard and Jan Mayen Islands	SJM
Sweden	SWE
Switzerland	CHE
Syria	SYR
Taiwan	TWN
Tajikistan	TZK
Tanzania	TZA
Tristan da Cunha, Ascension, and St. Helena Islands	SHN
Tunisia	TUN
Turkey	TUR
Turkmenistan	TKM
U.S.A.	USA
Uganda	UGA
Ukraine	UKR
United Kingdom	GBR
Uruguay	URY
Uzbekistan	UZB
Venezuela	VEN
Virgin Islands (U.S.A.)	VIR
Yemen	YEM
Yugoslavia	YUG
Zaire [Democratic Republic of Congo]	ZAR

SYNOPSIS OF THE GENUS *GRIMMIA*

1. **Grimmia alpestris** (F. Weber & D. Mohr) Schleich., Cat. Pl. Helv. ed. 2: 29. 1808. *Grimmia alpestris* Schleich., Neues J. Bot. 1(2): 196. 1805, nom. nud. *Trichostomum pulvinatum* var. *alpestre* Schleich. ex F. Weber & D. Mohr, Bot. Taschenbuch: 110. 1807. *Campylopus pulvinatus* var. *alpestris* (F. Weber & D. Mohr) Brid., Muscol. Recent. Suppl. 4: 75. 1818 ["1819"]. *Dryptodon pulvinatus* var. *alpestris* (F. Weber & D. Mohr) Brid., Bryol. Univ. 1(1): 198. 1826. *Grimmia donniana* var. *sudetica* Huebener, Muscol. Germ.: 175. 1833, nom. illeg., incl. typ. var. prior. *Grimmia donniana* var. *alpestris* (F. Weber & D. Mohr) Hampe, Flora 20: 281. 1837. *Guembelia alpestris* (F. Weber & D. Mohr) Hampe, Bot. Zeitung (Berlin) 4(8): 125. 1846. *Grimmia alpestris* var. *eualpestris* Loeske, Laubm. Eur. Part I: 101, fig. 25b, 27a–c. 1913, nom. inval. PROTOLOGUE: [Switzerland] Grimm. alpestris Schleich. Cent. 2, n° 13 (in monte Sylvio ad rupes a Schleichero lecta) et Dicr. piliferum ej. l. c. 4. n. 12. (in calidiorr, Vallesiae ad saxa inventum [...] In subalp., in rup. (in summo Bruct. ipsi legimus). TYPE: In M. Sylvio ad rupes (lectotype, designated by Muñoz (1997), BM!).

Grimmia holzingeri Cardot & Thér., in Holz., Bot. Gaz. 30: 123, tab. 11 fig. 2. 1900 [Wijk et al. (1962: 388) erroneously cited volume 15, 1890]. *Grimmia alpestris* var. *holzingeri* (Cardot & Thér.) G. N. Jones, in Grout, Moss Fl. N. Amer. 2: 31, pl. 9. 1933. *Grimmia donniana* var. *holzingeri* (Cardot & Thér.) Wijk & Margad., Taxon 9: 190. 1960. PROTOLOGUE: [U.S.A., northwestern Montana, leg. J.M. Holzinger & J. Blake, July 1898] Base of Sperry glacier; Mt. Trilby. TYPE: [...] vicinity of Lake McDonald, Flathead county, N.-W. Montana [...] four miles north of Belton, a station 30 miles east of Kalispell, on the Great Northern Railway. Mt. Trilby. Collected by John M. Holzinger and James B. Blake. July 21. 1898. 3,500–7,000 Feet (lectotype, designated by Muñoz (1998b: 384), PC!; isolectotypes, FH!, H-BR! 2 replicates, NY! 2 replicates, PC!). Synonymized by Muñoz (1998b: 384).

Grimmia alpestris var. *breviseta* J.J. Amann, Fl. Mouss. Suisse 2: 139. 1918. PROTOLOGUE: [Switzerland] Sur le gneiss erratique à Davos, 1600 m. (Amann) (BH 87.4.26). TYPE. Villa Amstein, Davos 1650–1700, 20.4.90. Leg. Amann (BH 87.4.26) (holotype, Z!).

DISTRIBUTION: AFG (*M. Köie s.n.*, MA); AND (*A. Sotiaux s.n.*, BCB); ARM (*V. Manakyan 4245*, ERE); AUT (*C.C. Townsend 92/522*, MA); BGR (*J. Podpěra s.n.*, TRH); CAN (*J. Macoun s.n.*, S); CHE (*J. Amann BH-18*, Z); CZE (*K.G. Limpricht s.n.*, BP); DEU (*Anonymous s.n.*, TRH); ESP (*J. Muñoz s.n.*, MA); FRA (*R.B. Pierrot 50377*, PC); GBR (*H.H. Knight s.n.*, FH); GEO (*P.H. Raven 27128B*, MO); ITA (*E. Corti s.n.*, FH); KAZ (*B. Allen 10535*, MO); NOR (*N. Bryhn s.n.*, TRH); NPL (*C. Troll 67*, JE); PAK (*K.H. Rechinger s.n.*, W); POL (*J. Wójcicki s.n.*, MA); ROM (*A. Boros s.n.*, BP); SFE (*I.V. Chernyadjeva 16*, LE); SIB (*M. Ignatov 0/477B*, MO); SVK (*T. Chałubinski s.n.*, BP); TUR (*Davis & Hedge D.30403*, MA); TZK (*U.K. Mamatkulov s.n.*, LE); UKR (*T. Alexeenko s.n.*, LE); USA (*A.T. Whittemore 3031*, MO); UZB (*A. Regel s.n.*, H-BR).

2. **Grimmia americana** E.B. Bartram, Bryologist 32: 8, pl. 1. 1929. PROTOLOGUE: [U.S.A.] Fort Davis, Jeff Davis Co., Texas, altitude 5200 feet, May 22–25, 1926, C.R. Orcutt, no. 7082. TYPE: Mosses of Western Texas. N° 7082. Fort Davis, Jeff Davis Co. alt. 5200 ft. C.R. Orcutt coll. May 22–25 1926 (lectotype, designated by Muñoz (1999a: 123), FH!; isolectotype, NY!).

DISTRIBUTION: USA (*C.R. Orcutt 7082*, FH).

3. Grimmia anodon Bruch & Schimp., Bryol. Europ. (fasc. 25–28) 3: 110, tab. 236. 1845. *Anodon ventricosus* Rabenh., Deutschl. Krypt.-Fl. 2(3): 154. 1848. *Schistidium anodon* (Bruch & Schimp.) Loeske, Laubm. Eur. Part I: 49. 1913. PROTOLOGUE: In muris et ad rupes calcareas: prope Heidelberg [Germany] ad arcem (Al. Braun), in Monte pinifero prope Gefrees (Funk); loco Bhühl prope Vindobonam [Austria] ad latera viae cavae (Dr. Putterlick); prope Mende Galliae [France] meridionalis (Prost); in alpibus Tyrolis et Styriae [Austria] (W.P. Sch.). TYPE: An Mauern bei Heidelberg. Al. Braun (lectotype, designated by Cao & Vitt (1986: 203), BM!; syntype ["Fichtelgebirge. *Funck*"], BM!).

Gymnostomum pulvinatum Hedw., Sp. Musc. Frond.: 36, tab. 3 fig. 1–7. 1801. *Anictangium pulvinatum* (Hedw.) D. Mohr, Ann. Bot. (König & Sims) 2: 545. 1806. *Anoectangium pulvinatum* (Hedw.) Röhl., Ann. Wetterauischen Ges. Gesammte Naturk. 2: 200. 1809. *Schistidium pulvinatum* (Hedw.) Brid., Muscol. Recent. Suppl. 4: 21. 1818 ["1819"]. *Anodon pulvinatus* (Hedw.) Rabenh., Deutschl. Krypt.-Fl. 2(3): 153. 1848. *Grimmia hoffmannii* Müll. Hal., Syn. Musc. Frond. 1: 780. 1849. *Grimmia sphaerica* Schimp., Coroll. Bryol. Eur.: 49. 1856, nom. illeg. incl. spec. prior. *Grimmia apocarpa* Hedw. subsp. *sphaerica* Boulay, Musc. France: 393. 1884. *Schistidium sphaericum* G. Roth, Eur. Laubm. 1: 398, tab. 30 fig. 5. 1904, nom. illeg. incl. spec. prior. *Schistidium apocarpum* (Hedw.) Bruch & Schimp. subsp. *pulvinatum* (Hedw.) Loeske, Biblioth. Bot. 101: 51, 61. 1930. *Grimmia apocarpa* Hedw. var. *pulvinata* (Hedw.) G.N. Jones, in Grout, Moss Fl. N. Amer. 2: 16. 1933. *Schistidium apocarpum* (Hedw.) Bruch & Schimp. var. *pulvinatum* (Hedw.) C.E.O. Jensen, in Weim., Förteckn. Skand. Växt., Moss. ed. 2: 28. 1937. PROTOLOGUE: [Germany] Ad pagum Reinhausen, prope Goettingam, legit misitque Hoffmann et Persoon. TYPE: Hab. ad rupestribus Göttingia leg. Hoffmann (lectotype, selected by Ochyra (1989: 94), LE not seen; isolectotypes, H-SOL, LE, MW, none seen).
Synonymized by Ochyra (1989a).

Grimmia alpina Kindb., Christiania Vidensk.-Selsk. Forh. 1888(6): 30. 1889. PROTOLOGUE: Reg. bet. in Jerkinhe prope viam veterem, Goverlivand versus! Reg. silv. super. Drivstuen!. TYPE: Norvegia: Dovre, Drivstuen. 3/7 87. N.C. Kindberg (lectotype, designated by Muñoz (1999a: 125), S!; isolectotype, S!; syntype ["Jerkinhe"], S!).
Synonymized by Muñoz (1999a: 125).

Grimmia anodon var. *sinaitica* Renauld & Cardot, Bull. Herb. Boissier 2: 33. 1894. PROTOLOGUE: [Egypt] Sinai, Djebel Senah. Leg. F. Grote, decemb. 12, 1891. TYPE: F. Grote. Herbier Sinaiticum. Djebel Senah. 12.XII.1891 (lectotype, here designated, PC!).

Grimmia dornajii Höhn., Sitzungsber. Kaiserl. Akad. Wiss., Math.-Naturwiss. Cl., Abt. 1, 104: 322. 1895. PROTOLOGUE: [Spain. Granada] Ich fand die Pflanze am 26.September 1892 in der Seehöhe von circa 1900 m auf Kalkfelsen, circa 200 m über Cortijo S. Geronimo am Westabhange des Dornajo [F.v. Höhnel]. TYPE: 26/9 1892. Kalkfelsen über Cortijo S. Geronimo c. 7000', Westseite des Dornajo, Sierra Nevada, Spanien. Leg. et det. Fr. v. Höhnel (lectotype, here designated, FH!).

Grimmia limprichtii Kern, Rev. Bryol. 24: 56. 1897. *Grimmia anodon* f. *epilosa* Mönk., Laubm. Eur.: 345. 1927. *Schistidium anodon* f. *limprichtii* (Kern) Loeske, Biblioth. Bot. 101: 74. 1930. *Schistidium anodon* f. *epilosum* (Mönk.) Podp., Consp. Musc. Eur.: 273. 1954 ["epilosa"]. PROTOLOGUE: [Switzerland] sur les rochers calcaires du limon della Palla, au-dessus du défilé de Rolle, dans les monts Dolomites, à une altitude de 2,100 mètres, le 29 juillet 1896 [F. Kern].

TYPE: Dolomiten, Palagruppe: Felsgallerien am limone, bei 2100m. 29.7.96 Kern (lectotype, here designated, FH!; isolectotypes, GOET!, JE!, PC!).

Grimmia edentula Hy, Rev. Bryol. 32(4): 83. 1905. PROTOLOGUE: [France] [...] Angers (Maine-et-Loire) sur le ciment calcaire d'un mur exposé au midi. TYPE: Murs du Chateau du Plain. 28 Janv. 1897 [F. Hy] (lectotype, here designated, ANGUC!).

Schistidium confertum var. *bohemicum* Vilh., Acta Bot. Bohem. 1: 53. 1922 ["bohemica"]. PROTOLOGUE: [Czech Republic] Habitat ad rupes in Bohemia occid.-merid. (Šumava) apud Spiesova skála prope Janovice (1919, leg. dr. Hilitzer). TYPE: Spiessova skála 184/9i9 2/VII 1919 leg. Hilitzer (holotype, Blom (1996: 262), PRC not seen).

Grimmia anodon var. *anomala* E.B. Bartram, Bryologist 27: 59. 1924. *Grimmia anodon* f. *anomala* (E.B. Bartram) G.N. Jones, in Grout, Moss Fl. N. Amer. 2(1): 22. 1933. PROTOLOGUE: [U.S.A.] On shaded banks of dry washes, Tucson Mts., Pima County, Arizona. E.B. Bartram, No. 931, February 13, 1924. TYPE: Mosses of Southern Arizona. Steep shaded banks of dry washes in Tuscon Mountains, Pima County. Altitude 2300 ft. Feb. 13 1924. Collected by Edwin B. Bartram (lectotype, designated by Muñoz (1999a: 125), FH!; isolectotype, H-BR!).
Synonymized by Muñoz (1999a: 125).

Schistidium tibetanum J.S. Lou & P.C. Wu, Acta Phytotax. Sin. 18: 121, pl. 2 fig. 8–15. 1980. PROTOLOGUE: [Tibet] Xizang: Baingoin on rocks, 10.VI.1976, B.S. Li 7601 (Type in Herb., Institute of Botany, Academia Sinica). TYPE: (holotype, PE not seen).
Synonymized by Cao & Vitt (1986: 203).

Schistidium obtusifolium Ireland & Crum, Bryologist 87: 371–373. 1984. *Grimmia subanodon* Ochyra, Mem. New York Bot. Gard. 45: 607–614. 1987. PROTOLOGUE: [and TYPE.] Canada, Quebec, Parc du Bic, Cap Enragé, pointe Ouest, 48°22'N, 68°46'W, on dry, calcareous conglomerate, 15 June 1983, Zoladcski & Lutzoni 11201 (holotype, QEF!; isotype, CANM!).

DISTRIBUTION: ARM (*N.V. Chikovani s.n.*, LE); AUT (*J. Fiedler s.n.*, H); BOL (*M. Lewis 79-1947*, F); CAN (*D.H. Vitt 23986*, H); CHE (*J. Amann BH-10*, Z); CHL (*H. Deguchi 26028*, HIRO); CHN (*B. Tan 95-199*, FH); CZE (*R. Vaněk s.n.*, MEL); DEU (*Schmidt 1952*, JE); DZA (*L. Trabut 1782*, PC); EGY (*A. Kneucker 13*, JE); ESP (*J. Muñoz s.n.*, MA); FRA (*J.L. De Sloover 34976*, MO); GEO (*A.L. Abramova & I.I. Abramov s.n.*, LE); GRC (*R.E. Magill 10074A*, MO); HUN (*R. Marstaller s.n.*, JE); IND (*W. Koelz 2898*, MO); IRN (*P. Goldblatt 830a*, MO); ITA (*F. Kern s.n.*, FH); JOR (*W. Frey & H. Kürschner 1136*, MA); KAZ (*B. Allen 10786*, MO); KGZ (*V.F. Brotherus s.n.*, FH); MAR (*J. Muñoz s.n.*, MA); MEX (*A.J. Sharp 3912*, TENN); NOR (*N.C. Kindberg s.n.*, S); PAK (*T. Thomson 287*, FH); PER (*H. Deguchi 29980*, HIRO); SAU (*W. Frey & H. Kürschner 2709*, MA); SFE (*O.M. Afonina s.n.*, LE); SIB (*M. Ignatov 0/478A*, MO); SUN (*V.P. Savicz s.n.*, FH); SVK (*I. Györffy 21*, H); SWE (*N. Hakelier s.n.*, MO); TKM (*I.V. Sirotina s.n.*, LE); TUR (*E. Nyholm & A.C. Crundwell 1229*, F); TZK (*T. Ahti 30324*, H); USA (*P.L. Redfearn Jr. 263-51*, MO); YUG (*M. Sabovljević s.n.*, MA).

4. Grimmia anomala Hampe ex Schimp., Syn. Musc. Eur. ed. 2: 270. 1876. *Grimmia elatior* subsp. *anomala* (Hampe) Kindb., Eur. N. Amer. Bryin. 2: 222. 1898. *Dryptodon anomalus* (Hampe) Loeske, Hedwigia 49: 32. 1909. *Grimmia hartmanii* var. *alpinoborealis* Loeske, Laubm. Eur. Part I: 141. 1913, nom. inval., pro synon. *Grimmia hartmanii* var. *anomala* (Hampe) Mönk., Laubm.

Eur.: 369. 1927. *Grimmia hartmanii* subsp. *anomala* (Hampe) Loeske, Biblioth. Bot. 101: 182. 1930. PROTOLOGUE: [Switzerland] ad rupes prope Zermatt Valesiae rever. Bertram legit. TYPE: (holotype, BM not seen).

Grimmia philibertiana E. Britton, Bull. Torrey Bot. Club 18: 51, pl. 114 fig. 18–28. 1891. PROTOLOGUE: [U.S.A.] Rare on quartz croppings of metalliferons veins, 6,000 ft. alt., divide between Traille River and Independence Creek, Sept. 27, 1889 (219). TYPE: 6000ft. alt. Traille River divide [illegible] Independence. Coll. J.B. Leiberg Sept. 27th 89. 219 (holotype, NY!; isotype, PC!).
Synonymized by Deguchi (1978: 163).

Grimmia pachyneurula Müll. Hal. & Kindb., in Macoun, Cat. Canad. Pl., Musci: 65. 1892. *Grimmia apocarpa* subsp. *pachyneurula* (Müll. Hal. & Kindb.) Kindb., Eur. N. Amer. Bryin. 2: 234. 1898. *Schistidium pachyneurulum* (Müll. Hal. & Kindb.) Ochyra, Fragm. Florist. Geobot. 43: 106. 1998. PROTOLOGUE: [Canada] On rocks, near the Whirpool on the west side of the Columbia, at Revelstoke, B.C. May 7th, 1890. (Macoun.). TYPE: Grimmia apocarpa pachyneurula C.M. et Kindb. ad int. N. Amer., Brit. Col., Revelstoke. 7/5 90 J. Macoun (lectotype, designated by Blom (1996: 262), S not seen).
Synonymized by Blom (1996: 262).

Grimmia phyllantha Lindb. ex Broth., Acta Soc. Sci. Fenn. 19(12): 83. 1892. *Racomitrium phyllanthum* (Broth.) Warnst., Hedwigia 33: 306. 1913. PROTOLOGUE: [Georgia] Cauc. major occ.: Radscha, in monte Bereleti pr. Uzeri, ad saxa regionis alpinae. TYPE: Radscha, in alpe Bereleti pr. Uzeri, ad saxa reg. alp. 3/7/1877. Leg. V.F. Brotherus (holotype, H-BR!; isotypes, H-SOL!, PC! 2 replicates).
Synonymized by Deguchi (1978: 163).

DISTRIBUTION: AUT (*W.E. Nicholson & H.N. Dixon s.n.*, JE); CAN (*G.R. Brassard 10990b*, NY); CHE (*P. Culmann s.n.*, MA); ESP (*I. Aizpuru s.n.*, MA); GEO (*G. Woronow 114*, PC); GRL (*K. Holmen s.n.*, NY); IND (*L. Borelli s.n.*, H-BR); ISL (*B. Jóhannsson 11992*, ICEL); NOR (*Bryhn s.n.*, JE); SWE (*C. Stenholm s.n.*, MO); USA (*D.H. Norris 57676*, NY).

5. Grimmia arcuatifolia Kindb., Bull. Torrey Bot. Club 16: 93. 1889. PROTOLOGUE: [Canada] Abundant on dry rocks at Cedar Hill, near Victoria, Vancouver Island. April 21, 1887. John Macoun. TYPE: Canadian musci nº 608. On dry rocks, Mount Tolmie. Near Victoria, V.I., April 21st, 1887. Coll. J. Macoun (lectotype, designated by Muñoz (1999b: 7), CANM-198077!; isolectotype, FH!).

Racomitrium alternuatum Müll. Hal. & Kindb., Cat. Canad. Pl., Musci: 73. 1892. *Grimmia alternuata* (Müll. Hal. & Kindb.) Kindb., Eur. N. Amer. Bryin. 2: 228. 1898 ["attenuata"]. *Grimmia attenuata* (Müll. Hal. & Kindb.) Müll. Hal. & Kindb, Eur. N. Amer. Bryin. 2: 228. 1898, nom. inval., orthogr. err. PROTOLOGUE: [Canada] On boulders at the base of Avalanche Mountain, Roger's Pass, Selkirk Mountains, B.C., Aug. 5th, 1890. (Macoun). TYPE: along the C. P. R'y at the summit of Roger's Pass, Selkirk Mountains, Aug. 6th, 1890. [Macoun, Canadian musci, nº 524] (lectotype, designated by Muñoz (1999b: 7–8), FH!; NOTE: Other duplicates are *G. elatior* and are excluded from the type).

Grimmia procera Kindb., Rev. Bryol. 23: 18. 1896, nom. illeg., non Bals.-Criv. & De Not. 1838. PROTOLOGUE: Amer. Canada, Vancouver isl. (without capsules): Macoun. TYPE: On rocks, Parson's Mt. Van. Island. Col. Macoun. Date May 19th, 1893. [Macoun, Canadian mosses n° 88a, n° 133] (lectotype, designated by Muñoz (1999b: 8), CANM-198094!; isolectotypes, CANM-198095!, FH!, NY!).

DISTRIBUTION: CAN (*R.L. Halbert 4461*, FH); USA (*D.L. Branscomb & J. Grunman 22826*, MO).

6. Grimmia arenaria Hampe, Linnaea 10: 405. 1836. *Grimmia curvula* Bruch & Schimp., Bryol. Europ. (fasc. 25–28) 3: 113, tab. 238. 1845, nom. illeg., incl. typ. spec. prior. *Grimmia incurva* Schleich. ex Bruch & Schimp., Bryol. Europ. (fasc. 25–28) 3: 113. 1845, nom. inval., pro synon. *Grimmia donniana* var. *curvula* Spruce, Musci pyrenaicae n° 281. 1847. *Grimmia donniana* subsp. *arenaria* (Hampe) Dixon, Stud. Handb. Brit. Mosses ed. 2: 155. 1904. *Grimmia donniana* var. *arenaria* (Hampe) Loeske, Laubm. Eur. Part I: 93, fig. 1c, 17a, 26b, 28. 1913, nom. illeg., incl. typ. var. prior. PROTOLOGUE: [Germany] Habitat in rupibus et saxis arenosis ad arcis pedem Regenstein, aestate fructu maturo. TYPE: ad rup. aren. Regenstein Hercyn[iae]. Junis (lectotype, designated by Muñoz (1998b: 376), BM!; isolectotype, FH!).

DISTRIBUTION: AUT (*H. Gander s.n.*, BP); CHE (*C. Trautmann s.n.*, S); DEU (*E. Hampe s.n.*, BM); ESP (*C. Casas s.n.*, MA); FIN (*V.F. Brotherus s.n.*, H-SOL); FRA (*L.A. Brébisson s.n.*, FH); GBR (*D.A. Jones & P.G. Rhodes 2169*, BP); ITA (*V. de Cesati s.n.*, G).

7. Grimmia atrata Miel. ex Hornsch., Flora 2: 85. 1819. *Grimmia alpestris* var. *brevipila* Bruch ex Mont., Arch. Bot. 1: 219. 1833, nom. inval., pro synon. *Dryptodon atratus* (Hornsch.) Hartm., Handb. Skand. Fl. ed. 3: 271. 1838. PROTOLOGUE: [Austria] Salzburg, an Felsen bei der Grube Schwarzwand in der Grosarl. Iul. TYPE: Grosarl. Hornschuch misit 1819 (lectotype, designated by Deguchi (1978: 170), B!; isolectotype, F!).

Grimmia strictifolia Mitt., J. Linn. Soc., Bot. Suppl. 1: 44. 1859. *Racomitrium strictifolium* (Mitt.) A. Jaeger, Ber. Tätigk. St. Gallischen Naturwiss. Ges. 1872–73: 89 (Gen. Sp. Musc. 1: 371). 1874. PROTOLOGUE: [India] In Himalayae orient. reg. alp., Sikkim, J.D. Hooker! TYPE: 295. Jongri alt. 13.000 (lectotype, designated by Muñoz (1999a: 126–127), NY!).
Synonymized by Cao & Churchill (1995: 508).

Grimmia tristicha Herzog, Biblioth. Bot. 87: 58. 1916, nom. illeg., non Schwaegr. 1811. *Grimmia tristichoides* Margad., Lindbergia 1: 128. 1972, nom. nov. PROTOLOGUE: An feuchten Felsen im Hochtal von Viloco, ca. 4700 m, No. 3152 c. fr. u. 3170 (ster.); an nassen Felsen bei der Mine Chojñacota, ca. 4700 m, No. 2979. TYPE: Plantae in itinere secundo per Boliviam lectae, an nassen Felsen im Hochtal von Viloco. ca. 4700m. leg. Th. Herzog n° 3152. Oktober 1911 (lectotype, designated by Muñoz (1999a: 127), JE!; isolectotype, PC!; syntypes ["*Herzog 2979*," "*Herzog 3170*"], JE!, NY!).
Synonymized by Muñoz (1999a: 127).

Grimmia tristicha var. *comosa* Herzog, Biblioth. Bot. 87: 59. 1916. *Grimmia tristichoides* f. *comosa* (Herzog) Margad., Lindbergia 1: 128. 1972. PROTOLOGUE: An nassen Felsen im Hochtal von Viloco, ca. 4700 m, No. 3188. TYPE: Plantae in itinere secundo per Boliviam lectae. An nassen Felsen im Hochtal von Viloco. ca. 4700m. leg. Th. Herzog. Oktober 1911 (lectotype, designated by Muñoz (1999a: 127), JE!; isolectotype, PC!).
Synonymized by Muñoz (1999a: 127).

DISTRIBUTION: AUT (*H.C. Funck s.n.*, H); BEL (*J.-L. De Sloover 13701*, PC); BOL (*M. Lewis 87-884*, MO); CZE (*Z. Pilous s.n.*, H); DEU (*P.G. Lorentz s.n.*, PC); ESP (*P. Montserrat et al. s.n.*, BCB); FRA (*J. Cardot s.n.*, H); GBR (*D.G. Long & B.M. Murray 14767*, MA); IND (*J.D. Hooker 295*, NY); ITA (*Artaria s.n.*, JE); JPN (*Faurie 3380*, NY); NOR (*N. Hakelier s.n.*, MO); SWE (*O. Mårtensson s.n.*, MO); TWN (*S. Lin s.n.*, NICH)

8. Grimmia australis (Dixon & Sainsbury) J. Muñoz & Ochyra, J. Bryol., 21: 319. 1999. *Coscinodon australis* Dixon & Sainsbury, Trans. & Proc. Roy. Soc. New Zealand 75(2): 175. 1945. PROTOLOGUE: [New Zealand] Mount Cassidy, Arthur's Pass, Canterbury; coll. G. O. K. S. 31/12/1935, No. 827. TYPE: Mt. Cassidy, Arthur Pass, Canterbury, South Island, N.Z., 31/12/35. Coll. Sainsbury nº 827 (lectotype, designated by Muñoz & Ochyra (1999), BM!; isolectotypes, F!, FH!, NY!).

Grimmia immersa Ochyra, Fragm. Florist. Geobot. Suppl. 2, 1: 223. 1993. PROTOLOGUE: New Zealand. South Island. Canterbury. Mount Cook National Park, end of North Ridge, Mt. D'Archiac, elev. 6 100 ft., rotten rock and rubble; 26 January 1972, H.D. Wilson 2314 (Holotype: CHR-255623). TYPE: New Zealand. Mount Cook National Park. End of North ridge, Mt. D'Archiac 6100'. 26/I/1972. H. D. Wilson nº 2314 (holotype, CHR-255623!).
Synonymized by Muñoz & Ochyra (1999).

Grimmia perpilosa Ochyra, Fragm. Florist. Geobot. Suppl. 2, 1: 220. 1993. PROTOLOGUE: New Zealand. South Island. Southland. Fiordland National Park, Gertrude Saddle, c. 4 000 ft, on rock, 23.XII.74, John Child s.n. (holotype, BM!; isotype, KRAM). TYPE: Grimmia ? asperifolia c.fr. Gertrude Saddle. C. 4000'. 23.XII.74, on rock. John Child Collection (holotype, BM not seen; isotype, KRAM-99498!).
Synonymized by Muñoz & Ochyra (1999).

DISTRIBUTION: AUS (*J.M. Wilson s.n.*, MEL); NZL (*G.O.K. Sainsbury 827*, BM).

9. Grimmia austrofunalis Müll. Hal., Hedwigia 37: 165. 1898. PROTOLOGUE: Australia, Victoria, Grampians, Mt. William: Sullivan 1883 in Hb. Melbourne. TYPE: Grampians, mount William, 1883. Sullivan (holotype, MEL!; isotypes, MEL! 2 replicates, S!).
NOTE: A duplicate in H-BR is *G. trichophylla* and is excluded from the type.

Grimmia pansa R.S. Williams, Bull. New York Bot. Gard. 3(9): 125. 1903. *Grimmia pausa* R.S. Williams ex P. Syd., Just's Bot. Jahresber. 31(1): 262. 1904, nom. inval., orthogr. err. PROTOLOGUE: [Bolivia; leg.: R.S. Williams] Pelichuco, 3450 meters, May 4, 1902 (2823).

TYPE: Bolivia. Pelechuco, 3450 meters. May 4, 1902. R.S. Williams 2823 (lectotype, designated by Muñoz (1999a: 127), NY!; isolectotypes, FH!, JE!).
Synonymized with *G. trichophylla* by Deguchi (1987: 30), but see Muñoz (1999a: 127).

Racomitrium austrosudeticum Broth., in Herzog, Biblioth. Bot. 87: 60. 1916 ["Rhacomitrium"]. PROTOLOGUE: [Bolivia] Waldgrenze über Tablas, alt. 3400 m, No. 2859; Tunarisee, alt. 4400 m, No. 4807; Cerros de Malaga, alt. 4000 m, No. 4366; Choquecota chico, alt. 4500 m, No. 3179. TYPE: Waldgrenze über Tablas, 3400 m, Okt. 1911 Th. H. 2859 (lectotype, designated by Muñoz (1999a: 127), H!; isolectotype, JE!; syntypes ["*Herzog 3179*"], H!, JE!; ["*Herzog 4366*"], H!; ["*Herzog 4807*"], H!).
Synonymized by Muñoz (1999a: 127).

Grimmia crassiretis Cardot & Broth., Kongl. Svenska Vetenskapsakad. Handl. 63(10): 27, tab. 2 fig. 7. 1923. PROTOLOGUE: [Argentina] Patagonia andina: Lago San Martín (380); Valle Frias, Cerro Cáceres, in declivi meridionali, alt. c. 900 m.s.m. TYPE: 380. Patagonie: lac San Martín, forêt mixte. 27.1.1909. C. Skottsberg (lectotype, designated by Muñoz (1999a: 127), PC!; isolectotypes, H-BR!, UPS!; syntype ["Valle Frias"], PC!).
Synonymized with *G. trichophylla* by Deguchi (1984: 25), but see Muñoz (1999a: 127).

Grimmia benoistii Thér., Rev. Bryol. Lichénol. 9: 10. 1936. PROTOLOGUE: [Ecuador, leg. R. Benoist] Massif du Pichincha: rochers verticaux exposés au Nord, face au Condorguachana, 3.900 m. (n° 3116, 3117, 3114 p.p.). TYPE: 3116. en face de Condorguachana (massif du Pichincha) 30.X.1930. R. Benoist (lectotype, designated by Muñoz (1999a: 127), PC!; isolectotypes, F!, PC!; syntypes ["*Benoist 3117*"], PC!, F!).
Synonymized with *G. trichophylla* by Churchill (1994: 2), but see Muñoz (1999a: 127).

NOTE: All European specimens identified as *G. austrofunalis* are other species. Greven (1997: 830) listed several specimens, from which the first author was only able to study *Greven 2020* (S). It is typical *G. longirostris*, with reniform costa, long-rectangular, nodulose, juxtacostal cells, and no gemmae at costa back.

DISTRIBUTION: ARG (*C. Skottsberg 381*, PC); AUS (*C. Beauglehole 15019*, MEL); BOL (*M. Lewis 79-1354*, F); CHL (*H.H. Bartlett 19123*, NY); COL (*S.P. Churchill & W. Rengifo M. 17258*, MA); ECU (*R. Benoist 3116*, PC); NZL (*J. Child M1460*, F); PER (*J.-P. Frahm 824021*, MA); KEN (*E. Schelpe 12182*, MO); VEN (*D. Griffin III et al. PV-1170*, CAS).

10. Grimmia bicolor Herzog, Beih. Bot. Centralbl. 26(2): 66. 1909. PROTOLOGUE: Bolivia: Auf Felsblöcken im Llavetal (Prov. Cochabamba), ca. 3600 m; Januar, 08. TYPE: Flora von Bolivia. Auf Felsblöcken im Llavethal (Prov. Cochabamba) ca. 3600m. leg. Th. Herzog. Januar, 08 (lectotype, designated by Muñoz (1999a: 131), JE!; isolectotype, PC!).

Grimmia subquatricruris Broth., in Herzog, Biblioth. Bot. 87: 57. 1916. PROTOLOGUE: Felsen am Huaillattanisee, alt. 4900 m, No. 2973. TYPE: Bolivia, Felsen am Huaillattanisee, 4900 m,

10/1911. Leg. Th. Herzog 2973 (lectotype, designated by Muñoz (1999a: 131), H-BR!; isolectotype, JE!).
Synonymized by Muñoz (1999a: 131).

DISTRIBUTION: BOL (*M. Lewis 79-2581*, F).

11. Grimmia caespiticia (Brid.) Jur., Laubm.-Fl. Oesterr.-Ung.: 172. 1882. *Campylopus caespiticius* Brid., Muscol. Recent. Suppl. 4: 77. 1818 ["1819"]. *Dryptodon caespiticius* (Brid.) Brid., Bryol. Univ. 1(1): 200. 1826. *Grimmia funalis* var. *caespiticia* (Brid.) Hampe, Flora 20: 282. 1837 ["caespititium"]. *Guembelia caespiticia* (Brid.) Müll. Hal., Syn. Musc. Frond. 1: 773. 1849. *Guembelia caespitosa* Müll. Hal. ex Limpr., Laubm. Deutschl. 5(synon. reg.): 23. 1903, nom. inval., orthogr. err. *Grimmia cespiticia* Jur. ex Dalla Torre & Sarnth., Fl. Tirol 5: 268. 1904, nom. inval., orthogr. err. *Grimmia alpestris* var. *caespiticia* (Brid.) G.N. Jones, in Grout, Moss Fl. N. Amer 2: 30. 1933. PROTOLOGUE: [Switzerland] In Monte Sancti Bernhardi Helvetiae habitat, unde lectum accepi. TYPE: St. Bernard (holotype, B!).

Grimmia sulcata Saut., Flora 24: 39. 1841. *Guembelia sulcata* (Saut.) Hampe, Bot. Zeitung (Berlin) 4(8): 125. 1846. PROTOLOGUE: [Austria. Pinzgau...bei Annaberg] An Felsen, gleich unter dem Kamm des nördlichen Ausläufers des Sauecks im Norden von Stuhlfeden, neben dem Alpenwege, der in das Schwarzachenthal führt in 6000' Seehohe [...] August 1837. TYPE: Pinzgauer Alpen. Com. Dr. Sauter (lectotype, designated by Muñoz (1998b: 387), H-SOL!; isolectotype, Z!).
Synonymized by Muñoz (1998b: 387).

Grimmia jacquinii var. *subimberbis* Lindb., Öfvers. Förh. Kongl. Svenska Vetensk.-Akad. 23: 552. 1867. *Grimmia alpestris* var. *subimberbis* (Lindb.) Berggr., Kongl. Svenska Vetenskapsakad. Handl. 13(7): 49. 1875. *Grimmia caespiticia* var. *subimberbis* (Lindb.) Limpr., Laubm. Deutschl. 1: 780. 1889. *Grimmia sulcata* var. *subimberbis* (Lindb.) G. Roth, Eur. Laubm. 1: 431. 1903. *Grimmia caespiticia* f. *subimberbis* (Lindb.) Podp., Consp. Musc. Eur.: 283. 1954. PROTOLOGUE: [Norway. Spitsbergen] Amsterdamön. TYPE: Spitsb., Amsterdam Island. 1861. Leg. Holmgren (lectotype, designated by Muñoz (1998b: 387), H-SOL!).
Synonymized by Muñoz (1998b: 387).

Grimmia manniae Müll. Hal., Flora 70: 223. 1887. *Grimmia alpestris* var. *manniae* (Müll. Hal.) G. N. Jones, in Grout, Moss Fl. N. Amer. 2(1): 31, pl. 9. 1933. *Grimmia donniana* var. *manniae* (Müll. Hal.) Wijk & Margad., Taxon 9: 50. 1960. PROTOLOGUE: [U.S.A.] California, Napa Co., Napa Springs: Domina Martha Mann 2. Majo 1886 legit. TYPE: Napa Soda Springs. California. May 3. 1886. Martha R. Mann (lectotype, designated by Muñoz (1998b: 387), NY!; isolectotypes, FH!, PC!).
Synonymized by Muñoz (1998b: 387).

Grimmia pyrenaica Kern, Jahresber. Schles. Ges. Vaterl. Cult. 92: 38. 1915. PROTOLOGUE: [Spain] Auf eisenhaltigen Felsen auf der spanischen Seite des Port de Boucharo, c. 2230 m. TYPE: Gavernie, auf eisenhalbsigen Felsen dem Port de Boucharo, spanische Seite. 13-VII-1914. Kern (lectotype, designated by Bednarek-Ochyra et al. (1992: 392), WRSL!; isolectotypes, BP-80423!, JE!, KRAM!, MA-8158!, MA-10927!).
Synonymized by Bednarek-Ochyra et al. (1992: 392).

DISTRIBUTION: AND (*R.B. Pierrot 201/100*, PC); ARM (*V. Manakyan 4996a*, ERE); AUT (*J.-P. Frahm 921007*, MA); BEL (*A. Cornet s.n.*, NY); BGR (*A. Ganeva s.n.*, MO); CHE (*C. Bender BH-16*, Z); DEU (*Anonymous s.n.*, BP); ESP (*J. Muñoz s.n.*, BCB); FRA (*J.P. Hébrard 1611774*, PC); GEO (*G. Woronow 192A*, PC); ITA (*J.P. Hébrard 7-2882*, PC); POL (*T. Chałubinsky s.n.*, BP); PRT (*C. Sérgio & A. Séneca 8164*, MA); ROM (*L. Vajda s.n.*, BP); SIB (*L.V. Bardunov s.n.*, NICH); SJM (*H.J. Holmgren s.n.*, H-SOL); SUN (*A.H. Brotherus & V.F. Brotherus 22*, H-SOL); SVK (*Z. Pilous s.n.*, G); TUR (*K. Walther 3318*, NY); USA (*E. Lawton 4792*, MO).

12. Grimmia capillata De Not., Mem. Reale Accad. Sci. Torino 39: 248. 1836. *Grimmia crinita* var. *capillata* (De Not.) De Not., Syllab. Musc.: 240. 1838. PROTOLOGUE: Ad muros vetustos hortorum meridiem versus. TYPE: (RO not seen).

Grimmia mairei Cardot & Copp., Bull. Soc. Bot. France 58: 500, fig. 15a–m. 1911. PROTOLOGUE: [Algeria] [...] recueillie par M. René Maire, à qui elle est dédiée, sur les rochers de grès du ravin de Ben Zireg, dans l'Extrême-Sud Oranais. TYPE: Rochers de grès à Beni-Ounif, Sahara algérien. Leg. R. Maire - 9/4/06 (lectotype, here designated, PC!; isolectotype, PC!).

Grimmia mesopotamica Schiffn., Ann. K. K. Naturhist. Hofmus. 27: 488. 1913. PROTOLOGUE: [Iraq] Ad terram et rupes calciferas Mesopotamiae mediae, c.sp. Erde der Gipswüste unterhalb Hit am rechten Ufer des Euphrat, ca. 150 m, 10./IV.1910 (Nr. 844); Kalkmergelfelsen bei der Dorfruine Gharra in der Mitte des Nordfußes des Dschebel Abd-el-Asis, ca. 500 m, 21./VI.1910 (Nr. 1778). TYPE: No. 844. Ad confines Mesopotamiae et Arabia borealis: Ad Euphratem medium inter Abukemal et Ramadi, in desertis infra oppidum Hit. Substrato gipsaceo; ca. 120–180 m. Leg. 10.IV.1910 Dr. Heinr. Frh.v. Handel-Mazzetti. (Tageb.- No. 154) determ. Schiffner (lectotype, here designated, FH!; isolectotype, FH!; syntypes ["*Handel-Mazzetti 1778*"], FH!, H-BR!).

Grimmia capillata var. *calva* J.J. Amann, in E. Bauer, Musci europaei et americani exsiccati ser. 39 n° 1915. 1927. *Grimmia crinita* var. *calva* (J.J. Amann) J.J. Amann, in Loeske, Biblioth. Bot. 101: 88. 1930. PROTOLOGUE: Frankreich: Aix-en-Provence, an Mauern der Route d'Avignon, A. 18.Sept.1925 und B. 31.Mai.1926 legit J. Amann. TYPE: Frankreich: Aix-en-Provence, an Mauern der Route d'Avignon, B. 31.Mai.1926 legit J. Amann (lectotype, here designated, MA-7092!; isolectotype, JE!).

DISTRIBUTION: DZA (*L. Trabut 1761*, PC); ESP (*J. Guerra et al. s.n.*, MA); IRQ (*H.F.v. Handel-Mazzetti 844*, FH); ISR (*I. Herrnstadt 4510*, FH); JOR (*U. Baierle & H. Kürschner 2579*, MA); MAR (*J. Muñoz s.n.*, MA); PAK (*C. Troll 38*, JE); SIB (*L. Bardunov s.n.*, ALTA); SYR (*H.F.v. Handel-Mazzetti 1778*, FH); TKM (*M.M. Il'in s.n.*, LE); TUN (*T.P. Maslin B-32420*, NY).

13. Grimmia crinita Brid., nom. conserv. prop., Muscol. Recent. Suppl. 1: 95. 1806. *Guembelia crinita* (Brid.) Hampe, Bot. Zeitung (Berlin) 4(8): 124. 1846. PROTOLOGUE: In totâ Europâ Australiori habitat miseram vitam in aridis degens. In Pago Vaudensi ad muros Lausannae legi, Roger Neoduni; praeterea in Valle-Clausa Gallo Provinciae, Massiliae, in Occitania et Catalonia ad muros faxaque copiose reperi. A Januario ad Aprilem usque cum capsulis maturescentibus, perfectis essoetisque vidi. TYPE: Avril 02. Rogeri s.n. (lectotype, designated by Geissler & Maier (1995: 499), B!).

NOTE: The name *G. crinita* has been proposed for conservation (Muñoz, 2000b).

Dicranum phascoideum P. Beauv., Prodr. Aethéogam.: 54. 1805 ["phascoidum"]. PROTOLOGUE: [France] [...] M. Dufour, se trouve sur un des murs du bois de Boulogne, près Auteuil. TYPE: Herbier Palisot de Beauvois. Type de "Dicranum ? phascoideum". Sur les murs du bois de Boulogne. Echantillon original provenant de l'herbier Desportes (lectotype, designated by Muñoz (2000b), G!; isolectotype, PC-P.Beauv!).

Gymnostomum decipiens F. Weber & D. Mohr, Bot. Taschenbuch: 79. 1807. PROTOLOGUE: In saxis pr. Herbipolin [Würzburg]. (Eundem muscum ad fontem Vaucluse Galliae lect. mis. Bridel). TYPE: [Germany] Ex herb. D. T. [D. Turner]. Gymn. decipiens n. sp. Fl. Germ. Wirceburgi [Würzburg] (lectotype, here designated, BM!).

Grimmia crinita var. *libani* Bizot, Rev. Bryol. Lichénol. 13: 51. 1942. PROTOLOGUE: [Lebanon] Katana [à 25 km. au sud-ouest de Damas. Maurize Nickles, 1932]. TYPE: Katana, Beyrouth, Liban. 4 septembre 1932. leg. Nickles (holotype, PC!).

DISTRIBUTION: AZE (*Alexeenko s.n.*, LE); BEL (*C. Delogne s.n.*, FH); CHE (*P. Culmann s.n.*, MO); CZE (*Velenovský s.n.*, JE); DEU (*L. Meinunger 17843*, JE); DZA (*L. Trabut s.n.*, PC); ESP (*J. Muñoz s.n.*, MA); FRA (*O. Aicardi 501*, PC); GEO (*N.V. Chikovani s.n.*, LE); IRL (*S.W. Greene s.n.*, TCD); ISR (*F. Bilewsky 658*, PC); ITA (*Loitlesberger s.n.*, JE); LBN (*M. Nickles s.n.*, PC); TKM (*V.F. Brotherus s.n.*, FH-SULL); TUN (*P. Robinson s.n.*, MO).

14. Grimmia curviseta Bouman, J. Bryol. 16: 379. 1991. PROTOLOGUE: Tenerife (Spain, Canary Islands) Las Canadas [Cañadas] del Teide, Degollada de Guajara, 2400 m altitude, 3 January 1989, UTM: CS4020. Leg. G.M. Dirkse RIN 005828. The holotype is preserved at U (isotype at L). TYPE: (holotype, U not seen).

DISTRIBUTION: ESP (*A.C. Bouman 91169*, NY!).

15. Grimmia decipiens (Schultz) Lindb., in Hartm., Handb. Skand. Fl. ed. 8: 386. 1861. *Trichostomum decipiens* Schultz, Prodr. Fl. Starg. Suppl.: 70. 1819. *Dryptodon schultzii* Brid., Bryol. Univ. 1(1): 199. 1826, nom. illeg., incl. typ. spec. prior. *Dryptodon grimmioides* Brid., Bryol. Univ. 1(1): 199. 1826, nom. inval., pro synon. *Grimmia schultzii* Huebener, Muscol. Germ.: 195. 1833, nom. illeg., incl. typ. spec. prior. *Dryptodon decipiens* (Schultz) Loeske, Stud. Morph. Syst. Laubm.: 111. 1910. PROTOLOGUE: [Germany] Habitat ad saxa granitosa rarius. - ad pagum Gevezien an der Landstrafse. Junio c. operculo. TYPE: ad saxa granitosa Duc[ato] Megapol. Stargard (lectotype, designated by Geissler & Maier (1995: 499), G!; isolectotype, H-SOL!).

Dryptodon funalis var. *longipilus* Brid., Bryol. Univ. 1: 771. 1827. *Dryptodon longipilus* Brid., Bryol. Univ. 1: 771. 1827, nom. inval., pro synon. PROTOLOGUE: [Germany] In Hercynia Suevica passim ad rupes legit et communicavit D. Braun. In monte Vogeso Mougeot et Nestler etiam invenerunt. TYPE: In Hercynia Suevica [et Vogesis ad saxa madida [illegible]]. Braun 1825 (lectotype, here designated, B!).

DISTRIBUTION: ARM (*B. Dildaryan s.n.*, LE); BEL (*C. Delogne s.n.*, FH); CHE (*Schlephacke s.n.*, JE); DEU (*G. Roth s.n.*, JE); ESP (*J.L. Rupidera 29*, SALA); FRA (*F.G. Kneiff & C.P.W. Maercker s.n.*, FH); GBR (*Jones s.n.*, JE); GRC (*R. Düll 13*, GOET); IRL (*H.W. Lett s.n.*, TCD); ITA (*E. Levier s.n.*, NY); MAR (*M. Brugués s.n.*, BCB); MKD (*T. Herzog 570A*, JE); NOR (*H. Winter s.n.*, JE); POL (*K.G. Limpricht s.n.*, FH); PRT (*C. Sérgio et al. 30.1a*, MA); SWE (*O.J. Hasslow s.n.*, NY); TUR (*K. Walther 2407*, NY); UKR (*L. Partyka s.n.*, MA).

16. **Grimmia donniana** Sm., Engl. Bot. 18: 1259. 1804. *Grimmia sudetica* Schwägr., Sp. Musc. Frond. Suppl. 1(1): 87, tab. 24. 1811, nom. illeg., incl. typ. spec. prior. *Weissia donniana* (Sm.) Poir., in Lam., Encycl. Suppl. 5: 502. 1817. *Grimmia donnii* Gray, Nat. Arr. Brit. Pl. 1: 728. 1821, nom. illeg., incl. typ. spec. prior. *Dryptodon donnianus* (Sm.) Hartm., Handb. Skand. Fl. ed. 3: 270. 1838. *Grimmia obtusa* var. *donniana* (Sm.) Hartm., Handb. Skand. Fl. ed. 5: 377. 1849, nom. illeg., prior. ut spec. *Grimmia donii* Sm. ex Lindb., Musci Scand.: 30. 1879, nom. illeg. *Grimmia donniana* var. *eudonniana* Loeske, Laubm. Eur. Part I: 91. 1913, nom. inval. *Guembelia donniana* (Sm.) Loeske, Laubm. Eur. Part I: 90. 1913, nom. inval., pro synon. *Grimmia donniana* f. *sudetica* (Schkuhr) Loeske, Biblioth. Bot. 101: 113. 1930, nom. illeg., non Chal. 1882. *Grimmia donniana* subsp. *eudonniana* Giacom., Atti Ist. Bot. Lab. Crittog. Univ. Pavia, ser. 5, 4: 221. 1947, nom. inval. PROTOLOGUE: [United Kingdom] George Donn [...] gathered it on large stones near a water-fall on a mountain in Angusshire, 18 miles to the north of Forfar. Mr. Turner has also found it on old pales at Beddgelart, North Wales, and from his specimen we have delineated it. TYPE: North Wales, pales at Beddgelart. July 1802. D. Turner (lectotype, designated by Muñoz (1998b: 378), BM!).

Grimmia sudetica Spreng. ex Schkuhr, Deutschl. Krypt. Gew. Heft 2: 48, tab. 22. 1811. PROTOLOGUE: [Poland/Czech Republic border] Ludwig kryptog. Gewächse, cent. 1, n. 17. Habitat in saxis Sudet. TYPE: Ludwig Kryptogamische Gewäsche der Riesengebirgen. Laubmoose [handwritten, not the original label of Ludwig's exsiccata] (lectotype, designated by Muñoz (1998b: 378–379), M!; isolectotype, JE!).
Synonymized by Muñoz (1998b: 378).

Grimmia obtusa Schwägr., Sp. Musc. Frond. Suppl. 1(1): 88, tab. 25. 1811, nom. illeg., non Brid. 1801. *Grimmia donniana* var. *obtusa* Steud., Nomencl. Bot. 2: 189. 1824. *Dryptodon erostris* Hartm., Handb. Skand. Fl. ed. 4: 374. 1843, nom. illeg., incl. typ. spec. prior. PROTOLOGUE: [Austria] Legi in Alpe Carinthiae Breenkogel, summo monti Glockner vicina, Augusto fructificantem. TYPE: Glockner. Kaulfuss (holotype, G!; isotype, G!).
Synonymized by Deguchi (1978: 182).

Grimmia bohemica Schkuhr ex Steud., Nomencl. Bot. 2: 189 [also 188]. 1824, nom. inval., pro synon. *Grimmia donniana* var. *bohemica* Schkuhr ex Steud., Nomencl. Bot. 2: 189. 1824, nom. nud. *Grimmia donniana* var. *bohemica* Schkuhr ex Brid., Bryol. Univ. 1(1): 176. 1826. PROTOLOGUE: In sudetis. Ludw[ig, C.G.]. TYPE: [Poland] Ludwigius dedit 1814. Circa dem Schneekoppe Sudetum (lectotype, designated by Muñoz (1998b: 379), B!).
Synonymized by Muñoz (1998b: 379).

Grimmia obtusa var. *elongata* Nees & Hornsch., Bryol. Germ. 2(1): 137, tab. XX fig. 11b. 1827. *Grimmia obtusa* var. *mammilaris* Huebener, Muscol. Germ.: 177. 1833, nom. illeg., incl. typ. var. prior. *Grimmia donniana* var. *elongata* (Nees & Hornsch.) Wilson, Bryol. Brit.: 159. 1855. *Grimmia donniana* f. *elongata* (Nees & Hornsch.) Loeske, Biblioth. Bot. 101: 113. 1930. PROTOLOGUE: [Austria] Wächst am Wege von Gastein nach Böckstein auf Steinen, die, nachdem man das Wäldchen hinter sich hat, an der Acha liegen (Funck! Laurer! Mielichhofer! Hsch.), dem Wege zum Rathhausberg, wenn man die Alpenregion betritt, links auf Steinen (Funck! Laurer! Hsch.), auf den Judenburger Alpen (Hrsch. Willdenow!), auf dem Felbertauern, dem Wildbachkogel im Oezthal und dem Timpeljoch (Funck!). TYPE: Stein. Alp Hsch. Ex herb. Hsch. (lectotype, here designated, JE!; syntypes ["zwischen Gastein und Böckstein, *Funck*". Cryptogamische Gewächse besonders des Fichtelgebirges, fasc. 22, nº 450, 1815], H!, JE!, RO!).

Grimmia nivalis Liebm. ex Besch., Mexic. Pl. 1: 19. 1872, nom. nud. ORIGINAL MATERIAL: [Mexico] In monte Orizabensi, alt. 14,000 p. (Liebmann in herb. Schimper). ORIGINAL SPECIMEN: Pico d'Orizaba 14,000'. Liebmann leg. 42 (BM!).

Grimmia donniana f. *vulgaris* Chal., Pamietn. Fizyogr. 2: 56. 1882. *Grimmia donniana* f. *longiseta* Vilh., Věstn. Král. Ceské Společn. Nauk, Tř. Mat.-Přír. 1924: 17. 1924, nom. illeg., incl. typ. f. prior. PROTOLOGUE: [Poland] Zakopane: Chramcówki; wał kamienny wzdłuz Bystréj Wody; obficie. Krupówki; takiż wał. Dolina Małego Kolbachu wprost Łomnicy. Brzegi Morskiego Oka. Pod Polskim Grzebienem od strony Felki. TYPE: Dol. Małego Kolbachu 1.08.1878 [T. Chałubiński] (lectotype, here designated, ZAMU!).

Grimmia donniana f. *cylindrica* Podp., Sborn. Klubu Přír. Brně 5: 14. 1923. PROTOLOGUE: [Czech Republic] In summo monte Hokšár m. Sudetorum c.fr. 1911 VIII. coll. am. J. Schenk. TYPE: Gesenke: Hockschar. 1911 VIII leg. F. Schenk (lectotype, here designated, PR!).

DISTRIBUTION: AUT (*V. Schiffner s.n.*, BP); BOL (*M. Lewis 83-3243*, F); CHE (*J. Schleicher s.n.*, Z); CHN (*M. Zang 669*, ALTA); CZE (*J. Velenovský s.n.*, FH); DEU (*H.M.C. Solms-Laubach s.n.*, GOET); ECU (*P.J. Grubb 2501B*, FH); FRA (*N.J. Boulay s.n.*, FH); GBR (*D.G. Long & J.A. Paton 14015*, MA); GEO (*I. Abramov s.n.*, LE); GRL (*N. Hartz s.n.*, NY); HUN (*L. Vajda s.n.*, BP); IRL (*I. Carroll s.n.*, BM); ISL (*B. Jóhannsson 10828*, ICEL); ITA (*A. Carestia s.n.*, RO); JPN (*E. Ishiba 1331*, H-BR); MEX (*A.J. Sharp 4761a*, TENN); NOR (*C. Kaurin s.n.*, TRH); POL (*R. Ochyra s.n.*, BCB); ROM (*K. Demeter s.n.*, BP); SFE (*I.V. Chernyadjeva 37*, LE); SIB (*N. Zolotukhin s.n.*, MA); SVK (*A. Boros s.n.*, BP); SWE (*H. Smith s.n.*, TRH); UKR (*L. Partyka s.n.*, MA); USA (*F.J. Hermann 19583*, MO).

17. Grimmia elatior Bruch ex Bals.-Criv. & De Not., Mem. Reale Accad. Sci. Torino 40: 340. 1838. *Trichostomum incurvum* Hoppe & Hornsch., Flora 2: 89. 1819. *Dryptodon incurvus* (Hoppe & Hornsch.) Brid., Bryol. Univ. 1(1): 194. 1826. *Racomitrium incurvum* (Hoppe & Hornsch.) Huebener, Muscol. Germ.: 201. 1830. *Grimmia funalis* var. *incurva* (Hoppe & Hornsch.) Hampe, Flora 20: 282. 1837. *Dryptodon funalis* var. *incurvus* (Hoppe & Hornsch.) Hartm., Handb. Skand. Fl. ed. 3: 270. 1838. *Grimmia funalis* var. *robusta* De Not., Syllab. Musc.: 251. 1838. *Grimmia funalis* subsp. *elatior* (Bals.-Criv. & De Not.) Hartm., Handb. Skand. Fl. ed. 5: 376. 1849. *Dryptodon elatior* (Bals.-Criv. & De Not.) Loeske, Stud. Morph. Syst. Laubm.: 111. 1910. *Dryptodon funalis* var. *robustus* De Not. ex H. Möller, Ark. Bot. 26A(2): 105. 1934 ["robusta"], nom. inval. *Dryptodon funalis* var. *incurvatus* Hartm. ex H. Möller, Ark. Bot. 26A(2): 135. 1934, nom. inval., orthogr. err. PROTOLOGUE: [Austria] Auf kahlen Felsenblöcken bey Heiligenblut, mit vollkommenen Früchten im Juni, im Nafsfelde und am Wege von Mattrey nach Deferreggen ohne Früchte. TYPE: In rupibus zu Hgblut [Heiligenblut] (lectotype, designated by Cao & Vitt (1986: 201), GZU!).

Grimmia elatior f. *alpigena* J.E. Zetterst., Grimmiae et Andreaeae exsiccatae n° 15 a–b. 1861, nom. nud. ORIGINAL MATERIAL: [Norway] In saxis juxta Kongsvold alpium Dovrensium. 27/7/1858. J.E. Zetterstedt. [Zetterstedt, Grimmiae et Andreaeae exsiccatae, n° 15a]. In saxis juxta fluvium supra Rödsheim paroeciæ Lom, Norvegiæ mediæ. 30/6/1858. J.E. Zetterstedt. [ibidem, n° 15b] (H!).

Grimmia elatior f. *infra-alpina* J.E. Zetterst., Grimmiae et Andreaeae exsiccatae n° 14a–c. 1861, nom. nud. ORIGINAL MATERIAL: [Norway] In saxis montis Egeberg juxta Christianiam. 15/6/1858. J.E. Zetterstedt. [Zetterstedt, Grimmiae et Andreaeae exsiccatae, n° 14a]. In saxis littoreis juxta Stocklycke montis Omberg, Ostrogothiæ. Jun. 1855. J.E. Zetterstedt. [ibidem, n° 14b]. In rupibus juxta Gottsunda prope Upsaliam. 7/5//1858. J.E. Zetterstedt. [ibidem, n° 14c] (H!).

Racomitrium papillosum Warnst., Hedwigia 20: 166. 1881. *Grimmia papillosa* (Warnst.) Kindb., Bih. Kongl. Svenska Vetensk.-Akad. Handl. 7(9): 110. 1883. *Grimmia elatior* subsp. *papillosa* (Warnst.) Kindb., Eur. N. Amer. Bryin. 2: 222. 1898. PROTOLOGUE: Norwegen, Dovre, Kongsvold von N.C. Kindberg im Juli 1879 entdeckt. TYPE: Dovre, f. om Sprenbäcken. 14/7 79. N.C. Kindberg (lectotype, here designated, S!; isolectotype, JE!).

Grimmia aspera Müll. Hal., Nuovo Giorn. Bot. Ital., n.s. 4: 261. 1897. PROTOLOGUE: China interior, prov. Schen-si septentr., Hua-san, sterilis, Aug. 1895. TYPE: China interior, provincia Schen-si septentr. Hua-san. 25 Aug. 1895 legit Rev. Jos. Giraldi, determ. C. Müller n. 1448 (lectotype, designated by Cao & Vitt (1986: 201), H; isolectotypes, BM, PC! 2 replicates, S).
Synonymized by Cao & Vitt (1986: 201).

Grimmia cognata Cardot & Thér., Bot. Gaz. 37: 368, tab. 19 fig. 1. 1904. PROTOLOGUE: [U.S.A.] Colorado: along the Cogwheel Railway to Pike's Peak, alt. 2100–3000 m (J.M. Holzinger, 1896). TYPE: Colorado [...] along the Cogg Wheel Railway to Pike's Peak. June 4 and 5, 1896, by J.M. Holzinger. Altitude, 7,000–10,000 feet (lectotype, here designated, PC!; isolectotype, H-BR!).

Grimmia grandis Kindb., Rev. Bryol. 32: 36. 1905. PROTOLOGUE: 2. Found by J. Macoun 1904 in British Columbia, 5200–7500 feet above the sea. TYPE: Canada, B.C. lake Louise, rocks. 6000 ft. 13/9 04. Macoun (holotype, S!; isotypes, FH!, S!).
Synonymized by Allen (1995: 161).

Grimmia elatior var. *hystrix* Podp., Věstn. Klubu Přír. v Prostějově 8: 36. 1906. PROTOLOGUE: [Czech Republic, Moravia] Na skalách Kotliny na Jeseníku ve výši 1200m [on the rocks of the Kotlina in the Ash Mountains, 1200m]. TYPE: Kotlina na Jeseníku. 1904 VII leg. Dr. Podpěra (lectotype, here designated, PR!).

Grimmia madagassa Renauld & Cardot, Prodr. Fl. Bryol. Madagascar. Supplément. Essai Leucoloma: 103. 1909. PROTOLOGUE: Madagascar: plateau central (Hb. Mitten). TYPE: Herb. Mitten. Central Madagascar (lectotype, here designated, PC!).

NOTE: No type material was found in NY.

Grimmia elatior f. *epilosa* H. Möller, Musci europaei et americani exsiccati ser. 44 n° 2163. 1932. PROTOLOGUE and TYPE: Schweden: [Norrbottens:] Lule Lappmark, Jokkmokk, Akkatsforsen, 1. Juli 1921 legit Hj. Möller (lectotype, here designated, S!; isolectotype, S!).

Grimmia elatior f. *nigrescens* H. Möller, Ark. Bot. 26A(2): 108. 1934. PROTOLOGUE: [Sweden] Värmland. Sunne. Skarped fr. 1925 G. Aberg. TYPE: Värml. Sunne m. Skarped nich Baüna alp. 80 m. 7/9 1925. G. Aberg (lectotype, here designated, S!).

DISTRIBUTION: ARM (*A.A. Fedorov s.n.*, LE); AUT (*Laurer s.n.*, JE); CAN (*R.R. Ireland 14985*, G); CHE (*V. Schiffner s.n.*, FH); CHN (*C. Gao & T. Cao 16556*, ALTA); CZE (*J. Podpěra s.n.*, H); DEU (*Herzog s.n.*, FH); ESP (*C. Casas s.n.*, MA); FIN (*R. Oldberg s.n.*, H); FRA (*J.L. De Sloover 34438*, MO); GBR (*J. Fergusson s.n.*, NY); GEO (*V.F. Brotherus 240*, H-SOL); GRL (*K. Holmen 62-1665*, NY); IND (*C.C. Townsend 87/343*, herb. Whittemore); ITA (*E. Levier 656*, NY); KAZ (*B. Allen 10448*, MO); KGZ (*Y. Rakhmatulina s.n.*, LE); MDG (*Anonymous s.n.*, PC); MNG (*S.V. Maksimovich 785*, LE); NOR (*E. Adlerz s.n.*, S); POL (*K.G. Limpricht s.n.*, FH); SFE (*O.M. Afonina s.n.*, LE); SIB (*L.V. Bardunov s.n.*, NICH); SUN (*V.G. Onichenko 136/95A*, NY); SVK (*S. Lisowski s.n.*, H); SWE (*J.E. Zetterstedt s.n.*, H); USA (*E.B. Bartram 17913*, NY).

18. Grimmia elongata Kaulf., in Sturm, Deutschl. Fl. Abtheilung II, Cryptogamie 4(13): 24. 1812. *Dryptodon elongatus* (Kaulf.) Hartm., Handb. Skand. Fl. ed. 3: 271. 1838. PROTOLOGUE: [Austria] im August 1812 auf der Seethaler Alpe in Srenermarf in Steinrißen mit reifen Fruchten. TYPE: In Styria alpibus Kaulfuss (lectotype, designated by Geissler & Maier (1995: 500), B not seen).

Grimmia fuliginosa Schimp. ex Müll. Hal., Syn. Musc. Frond. 1: 778. 1849. *Schistidium fuliginosum* (Müll. Hal.) Ochyra, Fragm. Florist. Geobot. 43: 105. 1998. PROTOLOGUE: Mexico, in summo Pico de Orizaba, alt. 14000': Liebmann. TYPE: Pic d'Orizaba 14000'. Liebmann leg. & det. Fbri. 1844 (lectotype, designated by Muñoz (1999a: 134), BM!; isotypes, BM!, PC-Besch!). Synonymized by Muñoz (1999a: 134).

Grimmia orientalis Wilson, Hooker's J. Bot. Kew Gard. Misc. 9: 323. 1857, nom. nud. ORIGINAL MATERIAL: [India] 293. Sikkim-Himalaya temperata; Singalelah, alt. 11,000 ped., J.D.H. [J.D. Hooker]. ORIGINAL SPECIMEN: n° 293 Sikkim: Singalelah (FH!).

Grimmia vulcanica Besch., Ann. Sci. Nat. Bot., sér. 6, 9: 351. 1878. PROTOLOGUE: La Réunion: Cratère brùlant, P. Lepervanche, 1878. TYPE: La Réunion. Cratere brùlant. Lépervanche (lectotype, here designated, BM!; isolectotypes, JE!, PC-Besch!).

Grimmia andreaeacea Müll. Hal., Linnaea 43: 453. 1882. PROTOLOGUE: Argentinia subtropica, Cuesta de Caldeira, 27. Majo 1873. [leg. P.G. Lorentz]. TYPE: Cuesta de Caldeira, 27.V.1873 (lectotype, designated by Muñoz (1999a: 134), BM!; isolectotype, PC!).
Synonymized by Muñoz (1999a: 134).

Grimmia montana var. *epilosa* Grav. ex Delogne, Ann. Soc. Belge Microscop. 9: 21. 1885. *Grimmia montana* f. *epilosa* (Delogne) Podp., Consp. Musc. Eur.: 281. 1954. *Grimmia trichophylla* var. *epilosa* (Grav. ex Delogne) De Sloover, Bull. Jard. Bot. État 37: 445. 1967, nom. illeg., non Warnst. 1897. PROTOLOGUE: Willerzie (Gravet). TYPE: Willerzie (prov. Namur, Belgique) 14 September 1882. F. Gravet (lectotype, designated by Muñoz (1999a: 134), PC!; isolectotypes, JE!, PC!).
Synonymized by Muñoz (1999a: 134).

Grimmia decalvata Cardot, Bull. Herb. Boissier, sér. 2, 8: 332. 1908. PROTOLOGUE: Japon: Ontake, 2500–3200 m. (n. 3319, 3321, 3326); Komagatake, 2500 m. (n, 3316, 3342). TYPE: Japon: sommet du Komagatake, 2300 m. Leg. Faurie nº 3316 (lectotype, designated by Deguchi (1978: 185), PC!; isolectotypes, H-BR!, NY!, PC!).
Synonymized by Deguchi (1978: 185).

Grimmia tolucensis Cardot, Rev. Bryol. 38: 1. 1911. PROTOLOGUE: [Mexico] Etat de Mexico: volcan de Toluca (Hahn; Pringle, 1892, nº 27a). TYPE: C.G. Pringle Plantae mexicanae, 1892, nº 27a. State of Mexico, on rocks, volcano of Toluca, 1400 ft. 7-Sept. 1892 (lectotype, designated by Muñoz (1999a: 134), PC!; isolectotypes, FH!, JE!, PC! 2 replicates, TENN!).
Synonymized with *G. incurva* by Robinson & Hermann (1964: 173), but see Muñoz (1999a: 134).

Grimmia incurva var. *rhaetica* J.J. Amann, Fl. Mouss. Suisse 2: 133. 1918. *Grimmia incurva* f. *rhaetica* (J.J. Amann) Podp., Consp. Musc. Eur.: 284. 1954. PROTOLOGUE: [Switzerland] Graubünden: Scaletta, 2500 m (BH 89.1.4) (Amann). TYPE: Graübunden: Scalettapass, 2500, 1888. Leg. Amann (BH 87.1.4) (holotype, Z!).

Grimmia elongata var. *epilosa* J.J. Amann, Bull. Soc. Vaud. Sci. Nat. 54: 45. 1921, nom. illeg., non Matousch. 1903. PROTOLOGUE: Italie, Province de Côme, Monte Legnone, 1700 m. environ (Artaria) (BH. 90.5.21). TYPE: Italia. Como. Mt. Legnone, 1700 m, 29 VII 1897. Coll. F. A. Artaria (BH. 90.5.21) (holotype, Z!).

DISTRIBUTION: ARG (*P.G. Lorentz s.n.*, BM); AUT (*R.B. Pierrot 62115,* PC); BEL (*F. Gravet s.n.*, PC); BOL (*M. Lewis 83-3317*, F); BRA (*P. Dusén 611*, H-BR); CHE (*M. Anzi 128*, FH); CHN (*K. Lang 598*, ALTA); COL (*A.M. Cleef 2449*, COL); CZE (*Z. Pilous s.n.*, H); DEU (*Herzog s.n.*, JE); ESP (*M. Acón s.n.*, BCB); FRA (*R. Potier de la Varde s.n.,* PC); GBR (*C.H. Binstead s.n.*, H); GEO (*V.F. Brotherus 304*, H-SOL); GTM (*J.A. Steyermark 35548*, FH); IND (*J.D. Hooker 293*, FH); ISL (*B. Jóhannsson 33583*, ICEL); ITA (*F.A. Artaria s.n.*, Z); JPN (*E. Ishiba 1228*, H-BR); LSO (*J. van Rooy 3402*, NY); MEX (*A.J. Sharp et al. 1514*, TENN); NOR (*J.E. Zetterstedt s.n.*, H); PRT (*H. Persson s.n.,* PC); REU (*P. Lépervanche s.n.*, JE); ROM (*M. Péterfi s.n.*, BP); SIB (*H. Nilsson-Ehle s.n.*, H); SVK (*Z. Pilous s.n.*, H); SWE (*O. Mårtensson s.n.*, MO); TWN (*J. Hyvönen & S. Stenroos 3840*, MA); UGA (*G.H.S. Wood 1107a*, NY); VEN (*A. Schäfer-Verwimp & I. Verwimp 12280a*, MA); ZAF (*J. van Rooy 1114*, NY).

19. Grimmia funalis (Schwägr.) Bruch & Schimp., Bryol. Europ. (fasc. 25–28) 3: 119, tab. 247. 1845. *Trichostomum funale* Schwägr., Sp. Musc. Frond. Suppl. 1(1): 150, tab. 37. 1811. *Dryptodon funalis* (Schwägr.) Brid., Bryol. Univ. 1(1): 193. 1826. *Trichostomum patens* var. *funale* (Schwägr.) Brid., Bryol. Univ. 1(1): 193. 1826, nom. inval., pro synon. *Trichostomum patens* var. *piliferum* Hook. & Taylor, Muscol. Brit. ed. 2: 105. 1827. *Racomitrium funale* (Schwägr.) Huebener, Muscol. Germ.: 200. 1830. *Grimmia funalis* subsp. *eufunalis* Giacom., Atti Ist. Bot. Lab. Crittog. Univ. Pavia, ser. 5, 4: 222. 1947, nom. inval. *Grimmia funalis* var. *eufunalis* Podp., Consp. Musc. Eur.: 286. 1954, nom. inval. PROTOLOGUE: [Poland/Czech Republic border] Legit in subalpinis Sudeticis C. Ludwig. TYPE: Um die Schneekoppe L[udwi]g ex parte (lectotype, designated by Geissler & Maier (1995: 500), G!).

Grimmia spiralis Hook., in Grev., Scott. Crypt. Fl. 4: 203. 1826. *Dryptodon spiralis* (Hook.) Brid., Bryol. Univ. 1: 771. 1827. *Grimmia cernua* var. *spiralis* (Hook.) Nees & Hornsch., Bryol. Germ. 2(1): 177, tab. XXIV fig. 23a. 1827. *Grimmia incurva* var. *spiralis* (Hook.) Huebener, Muscol. Germ.: 191. 1833. *Grimmia uncinata* subsp. *spiralis* (Hook.) Hampe, Flora 20: 282. 1837. PROTOLOGUE: [United Kingdom] On rocks in the Scottish Highlands, at a considerable elevation. Throughout the whole or the Breadalbane range of mountains; in fruit upon Craig Challeach, Dr. Hooker, Mr. Walker Arnott, and Dr. Greville. Mountains of Clova, Mr. Drummond. TYPE: Craig Challeach July 1823 (Arnott) (lectotype, here designated, BM!; isolectotypes, PC! 3 replicates).

Grimmia incurva var. *flagellaris* Brid., Bryol. Univ. 1(1): 186. 1826. PROTOLOGUE: [Austria] In Heiligenbluter-Tauern Carinthiae superioris ad rupes sed semper absque fructu lectam misit clar. Hornschuchius. TYPE: Heiligenbluter Tauern. Hornsch. (lectotype, here designated, JE!).

Gymnostomum spirale Hartm., Handb. Skand. Fl. ed. 3: 265. 1838. PROTOLOGUE: [Norway] Fjellklipp. r. (N. Dovre: Blytt). TYPE: Dovre Kongsvold. Blytt (lectotype, here designated, S!; isolectotype, S!).

Grimmia funalis f. *epilifera* J.E. Zetterst., Revis. Grimm. Scand.: 74. 1861. *Grimmia spiralis* var. *epilifera* (J.E. Zetterst.) Hartm., Handb. Skand. Fl. ed. 9, 2: 56. 1864. *Grimmia imberbis* Kindb., Bot. Not. 1882: 186. 1882, nom. illeg., non Müll. Hal. 1849. *Grimmia calvescens* Kindb., Christiania Vidensk.-Selsk. Forh. 1888(6): 19. 1888, nom. nov. *Grimmia funalis* var. *epilifera* (J.E. Zetterst.) Broth. & Saelán, Acta Soc. Fauna Fl. Fenn. 6(4): 80. 1890. *Grimmia funalis* var. *epilosa* J.E. Zetterst. ex Paris, Index Bryol. 2: 526. 1895, nom. inval. *Grimmia tortifolia* var. *calvescens* (Kindb.) Kindb., in Röll, Hedwigia 35: 64. 1896. *Grimmia tortifolia* subsp. *calvescens* (Kindb.) Kindb., Eur. N. Amer. Bryin. 2: 217. 1898. *Grimmia funalis* subsp. *calvescens* (Kindb.) J.J. Amann, Fl. Mouss. Suisse 2: 137. 1918. *Grimmia bauri* Müll. Hal. ex H. Möller, Ark. Bot. 26A(2): 65. 1934, nom. inval., pro synon. *Grimmia funalis* var. *calvescens* (Kindb.) H. Möller, Ark. Bot. 26A(2): 65. 1934. PROTOLOGUE: Habitat in Norvegia media multis in locis, præcipue copiose in Dovre prope Kongsvold (ubi Kindberg perpauca specimina fructifera legit) et in monte Lauvhöen paroeciæ Lom; in Valders et in Voss eam legit Blytt. TYPE: Norvegia: Lom, Lauvhøen. 7/7/1858. J.E. Zetterstedt (lectotype, here designated, S!).

Grimmia ryanii Limpr. ex Bryhn, Nyt Mag. Naturvidensk. 34: 73, figs. 1–13. 1892 ["Ryani"]. *Grimmia tortifolia* subsp. *ryanii* (Bryhn) Kindb., Eur. N. Amer. Bryin. 2: 218. 1898. *Grimmia funalis* var. *ryanii* (Bryhn) Loeske, Biblioth. Bot. 101: 153. 1930. PROTOLOGUE: [Norway] August 1890 af Fabrikbestyrer Ryan ved Leirungsboden paa Südsiden af Gjendin i Jotunfjeldene

[...]. TYPE: Norvegia, Christians Amt. Jotunheimen, prope Leirungsboden in rupes siccis. Alt. metr. 1150. Aug. 1890. Chr. Kaurin & E. Ryan (lectotype, here designated, NY!; isolectotypes, JE!, PC!; syntypes ["Jotunfjeldene"], NY! 4 specimens).

Grimmia brittoniae R.S. Williams, Bull. Torrey Bot. Club 27: 316, pl. 19. 1900. PROTOLOGUE: [U.S.A.] Growing on shaded perpendicular walls, partly calcareous, in rather dry places. Collected for several seasons in Bad Rock Cañon, Flathead River, Mont. [Montana]. TYPE: nº 223. April 2, 1894. Montana. Columbia Falls. Collected by Robert Statham Williams (lectotype, designated by Muñoz (2000c), NY!; isolectotypes, NY! 3 replicates).
Synonymized by Muñoz (2000c).

Grimmia torquata var. *mutica* Herzog ex Loeske, Laubm. Eur. Part I: 159. 1913. *Grimmia torquata* f. *mutica* (Loeske) Loeske, Biblioth. Bot. 101: 159. 1930. PROTOLOGUE: [Switzerland] am "Seilhenker" des Medjekopfs in der Silvretta, +/- 2400 m. TYPE: Fl. Graubünden. An Gneisfelsen des Medjekopf (am Seilhenker). ca 2400 m. August, 1899 (lectotype, here designated, JE!).

Grimmia funalis var. *leptodictyon* Herzog, Kryptog. Forsch. 1: 292. 1919. PROTOLOGUE: [Romania] An schattig-feuchten Felsen einer Schlucht zwischen Poiana lunga und Fruntu, ca. 1000 m; an Felsen unter der Moscovulscharte, ca. 2000 m. TYPE: Transsilvan. Alpen. An Felsen einer Bachschlucht zwischen Poiana lunga u. Fruntu. ca 1000 m. November, 16 [1916]. T. Herzog (lectotype, here designated, JE!; syntype, ["Moscovulscharte, *Herzog s.n.*"] JE!).

Grimmia ryanii f. *tenuis* J.J. Amann, in Bauer, Musci Europaei et Americani Exsiccati, Ser. 43, nº 2215. 1931. *Grimmia funalis* f. *tenuis* (J.J. Amann) Podp., Consp. Musc. Eur.: 286. 1954. PROTOLOGUE and TYPE: Schweiz: Graubünden [...] 10. September 1930 legit J. Amann (lectotype, here designated, MO!; isolectotypes, JE!, NY!).

DISTRIBUTION: ARM (*V. Manakyan 4907a*, ERE); AUT (*J.-P. Frahm s.n.*, MA); AZE (*L. Lyubarkaya s.n.*, LE); BGR (*L. Meinunger 14486*, JE); CAN (*J.A. Hancock 1238*, NFLD); CHE (*P. Culmann s.n.*, H); CHN (*Jia Yu J02381*, MO); CZE (*Z. Pilous s.n.*, H); DEU (*L. Meinunger 4644,* JE); DZA (*L. Loeske s.n.*, JE); ESP (*C. Aedo s.n.*, MA); FIN (*V.F. Brotherus s.n.*, JE); FRA (*T. Husnot s.n.*, FH); FRO (*R. Fagerstén & M. Haapasaari 327*, H); GBR (*D.G. Long 6786*, MA); GEO (*A.H. Brotherus & V.F. Brotherus 41*, H-SOL); GRL (*K. Holmen 7436*, H); HUN (*A. Boros s.n.*, H); ISL (*B. Jóhannsson 11879*, ICEL); IRL (*H.W. Lett s.n.*, TCD); ITA (*M. Fleischer s.n.*, FH); MNG (*T. Tsegmed 331*, NY); NOR (*C. Kaurin & E. Ryan s.n.*, NY); POL (*R. Ochyra s.n.*, H); ROM (*A. Boros s.n.*, BP); SFE (*B.C. Tan 97-234*, FH); SIB (*M. Ignatov 0/469A*, MO); SUN (*V.G. Onichenko 136/95*, NY); SVK (*H. Zemanová s.n.*, H); SWE (*N. Hakelier s.n.*, MO); TUR (*Davis & Hedge D.32330*, MA); UKR (*K.O. Ulychna s.n.*, H); USA (*W.B. Schofield 12025*, FH); YUG (*M. Sabovljević s.n.*, MA).

20. Grimmia fuscolutea Hook., Musci Exot. 1: 63. 1818. PROTOLOGUE: In frigidis Tobeccae et Islahuacae, altitudine 1330 hexapod. (Regnum Mexicanum). Humboldt et Bonpland. TYPE: Nº 2673. S. Am. *Humboldt 50* (lectotype, here designated, BM!).

Grimmia apiculata Hornsch., Flora 1(19): 329. 1818, nom. nud. *Grimmia affinis* Hornsch., Flora 2(1): 85. 1819. *Grimmia apiculata* Hornsch., Flora 2(2): 442. 1819, nom. illeg., incl. typ. spec. prior. *Grimmia pulvinata* var. *apiculata* (Hornsch.) Huebener, Muscol. Germ.: 710. 1833, nom. inval., orthogr. err. *Dryptodon apiculatus* (Hornsch.) Hartm., Handb. Skand. Fl. ed. 3: 270. 1838.

PROTOLOGUE: [Austria] der Messerlingwand auf dem Windischmattreyer Tauern. TYPE: Windischmattreyer Tauern. Hornschuch missit (lectotype, designated by Deguchi (1978: 165), B!; isolectotype, H-SOL!).
Synonymized by Crum (1994: 403).

Grimmia abyssinica Bruch & Schimp., Bryol. Europ. (fasc. 25–28) 3: 109. 1845, nom. nud. *Guembelia abyssinica* Hampe, Bot. Zeitung (Berlin) 4(8): 125. 1846, nom. nud. *Guembelia abyssinica* Müll. Hal., Syn. Musc. Frond. 1: 772. 1849. *Grimmia abyssinica* (Müll. Hal.) Mitt., J. Linn. Soc., Bot. 7: 152. 1863. PROTOLOGUE: [Ethiopia] Abyssinia, monte alpestri Silke, 18. Febr. 1840: W. Schimper. TYPE: In siccis ad rupes inde a limitibus arbornm [sic] et fruticum usque ad cacumen montis Silke. d. 18.Febr.1840. U.i.[Unio Itineraria] 1842 (lectotype, here designated, MO!; isolectotypes: H-BR!, FH!, GOET!, JE!, NY!, PC!).

Grimmia holleri Molendo, Flora 50: 39. 1867. *Grimmia apiculata* var. *holleri* (Molendo) Husn., Muscol. Gall.: 133. 1887. *Grimmia apiculata* f. *holleri* (Molendo) Loeske, Laubm. Eur. Part I: 116. 1913. *Grimmia apiculata* subsp. *holleri* (Molendo) J.J. Amann, Fl. Mouss. Suisse 2: 132. 1918. PROTOLOGUE: [Austria. Salzburg: Alpen...] bei 5900' bis 6000' [...] (von der Stanz im Raufis-Fucher Scheiderlicken) 1864. TYPE: Rauris, in Menge am Rande d. Baumregion 6-6200', Glimmerschiefer. l. [legit] Holler. c. Mdo. (holotype, M!; isotype, M!).

Grimmia flexicaulis Müll. Hal., Linnaea 43: 454. 1882. PROTOLOGUE: Argentinia subtropica, Cuesta de Pinos, i.e., in descensu ab excelsa Puna in convallem Tarija Boliviae, 27. Martio 1873, [...]; Catamarca, supra Yakutula in convalle Granadillas in alpinis Vayas altas [sic], alt. 9–10,000 ped., in graniticis, initio Februarii 1872. [leg. P.G. Lorentz]. TYPE: Cuesta de Pinos: [illegible] Puna [illegible] Tarija in Bolivia. 27/3/73. Lorentz (lectotype, designated by Deguchi (1987: 22), JE!; isolectotype, BM!; syntypes, JE! ["Catamarca: Yakatula, Granadillas und Vayas"] 2 replicates).
Synonymized by Muñoz (1999a: 136).

Grimmia quatricruris Müll. Hal., Linnaea 43: 462. 1882. PROTOLOGUE and TYPE: Argentinia subtropica, Cuesta de Pinos in descensu ab alta Puna in convallem Tarija Boliviae, 27. Martio 1873. [leg. P.G. Lorentz] (lectotype, designated by Muñoz (1999a: 136), BM!; isolectotype, PC!).
Synonymized by Muñoz (1999a: 136).

Grimmia unicruris Müll. Hal., Linnaea 43: 463. 1882. PROTOLOGUE: Patria. Argentinia subtropica, Cuesta de Pinos in descensu ab alta Puna in convallem Tarija Boliviae, 27. Martio 1873,[...]. Quoque in Nevado de Castillo prope Salta; alpina, [...]. [leg. P.G. Lorentz]. TYPE: Cuesta de Pinos, 27.III.1873 (lectotype, designated by Muñoz (1999a: 136, 138), BM!).
Synonymized by Muñoz (1999a: 136).

Grimmia trichophylloidea Schimp. ex Müll. Hal., Nuovo Giorn. Bot. Ital., n.s. 4: 127. 1897. PROTOLOGUE: Bolivia, Llachissaxi, regione alpina 4200 m; Sorata, pr. Vaucuiri in scopulosis, 4500 m, Aprili 1860: Mandon legit. TYPE: Plantae Andium Boliviensium. Prov. Larecaja, vicinitis Sorata, prope Vancuiri, in scopulosis. Alt. Reg. alpina 4500. April 1860. G. Mandon nº 1633 (lectotype, designated by Muñoz (1999a: 138), NY!; isolectotypes, F!, PC!).
Synonymized by Muñoz (1999a: 138).

Grimmia glyphomitrioides Broth. ex Paris, Index Bryol. Suppl.: 173. 1900, nom. nud. ORIGINAL MATERIAL: Afr.: Cameroon. ORIGINAL SPECIMEN: Buea 3500 m (Cameroun). II 91. Preuss 859 (PC!).

Grimmia fastigiata Cardot, Bull. Herb. Boissier, sér. 2, 5: 1003. 1905. PROTOLOGUE: [Argentina] Terre-de-Feu: Ushuaia, mont Martial (Typus in herb. Cardot). TYPE: Svenska Südpolar-expeditionen 1901–03 n° 70. Terre-de-Feu: Ushuaia, mt. Martial, D. 11/3/1902 alt 1275 m. Carl Skottsberg. Det. J. Cardot (lectotype, designated by Muñoz (1999a: 138), PC!; isolectotypes, H-BR!, MEL-29537).
Synonymized by Muñoz (1999a: 138).

Grimmia afroincurva Broth., Wiss. Erg. Deut. Zentr.-Afr. Exped., Bot. 2: 148, tab. 13 fig. 17. 1910. PROTOLOGUE: [Zaire, District des Lacs Edouard et Kivu, Birunga Mountains]Vulkan-Gebiet: Karisimbi, Gipfelkegel, ca. 4000 m. ü. M. (reichlich fruchtend im Sept. 1907 - n. 2085). TYPE: Afr. or., Vulkan Gebiet, Karisimbi, 4000 m, 9/1907. Leg. J. Milbraed (lectotype, here designated, H-BR!).

Grimmia micropyxis Broth., Akad. Wiss. Wien Sitzungsber., Math.-Naturwiss. Kl., Abt. 1, 133(10): 567. 1924. PROTOLOGUE: [China] Prov. Setschwan austro-occid.: Cum praecedente [In montis Tschahungnyotscha trans flumen Yalung ad septentr. oppidi Yenyüen, 28°15', regione alpina, ad rupes schistosas; ca. 4150–4300 m] (Nr. 2671). TYPE: Prov. Setschwan austro-occid.: In montis Tschahungnyotscha supra vicum Ngaitschekou trans flumen Yalung ad septentr. oppidi Yenyüen ca. 28°15' siti regione alpina, ad rupes. Substr. schistaceo; alt. s.m. ca. 4150–4300 m. Leg. 27.V.1914 Dr. Heinr. Frh. v. Handel-Mazzetti (Diar. Nr. 498) n° 2671 (lectotype, here designated, W!; isolectotypes, JE!, PC!).
Synonymized by Deguchi (1978: 165).

Grimmia subdonniana Nog., Candollea 19: 175. 1964. PROTOLOGUE: Nepal: Moraine, near Lobuje (27°57', 86°49'), alt. 4870 m, April 2, 1952, Zimmermann 301-type, 304; moraine, near the base camp at Lobuje, alt. 5000 m, April 24, 1952, Zimmermann 325; near the base camp, in the direction of Lobuje, alt. 5100 m, May 31, 1952, Zimmermann 558. TYPE: Nepal. Moraine près de Lobuje, 4870 m. Apr. 2, 1952, coll. A. Zimmermann 301a. Ex Herb. Dr. A. Noguchi no. 68265 (holotype, NICH-365224!; paratype, G! [*Zimmermann 558*]).
Synonymized by Deguchi (1978: 165).

Grimmia sanii Greven, Bryologist 99: 429, fig. 2. 1996. PROTOLOGUE: South Africa. Drakensberg, Mount aux Sources, Crow's nest, alt. 13050 m [sic], Schelpe 2115 (holotype, PRE; isotypes, MO 2550366, MO 3679148). TYPE: South Africa. Natal: Drakensberg, Crow's Nest. Mount aux Sources. Alt. 10.000'. July 1947. E. A. Schelpe 2115 (isotypes, MO! 2 replicates).
Synonymized by Muñoz (1999a: 138).

DISTRIBUTION: ARG (*I.M. Lamb 5566*, FH); ATA (*R.I. Lewis-Smith 3677*, MA); AUT (*C.F. Hornschuch s.n.*, B); BOL (*M. Lewis 79-1824*, MO); CAN (*J.M. Macoun s.n.*, S); CHE (*T. Herzog s.n.*, JE); CHN (*C. Gao 1080*, ALTA); CMR (*P.R. Preuss 859a*, PC); COG (*H. Humbert 7430*, PC); ECU (*R. Benoist 3458*, F); ETH (*G.H.W. Schimper s.n.*, H-BR); IND (*J.F. Duthie s.n.*, H-BR); ITA (*C.C. Townsend 92/551*, MA); KEN (*O. Hedberg 950F*, UPS); LSO (*J. van Rooy 3475*, F); MEX (*Á. Cárdenas 3512*, ALTA); NOR (*N. Hakelier s.n.*, MO); NPL (*A. Zimmermann 304*, G); PER (*T.G. Tutin 196*, BM); REU (*P. Lepervanche s.n.*, PC); SIB (*M. Ignatov 0/476*, MO); SVK (*J. Śmarda s.n.*,

BP); TWN (*J. Hyvönen & S. Stenroos 3892*, MA); TZA (*O. Hedberg 2324*, PC); USA (*E.L. Little Jr. 10401*, MO); VEN (*A. Schäfer-Verwimp & I. Verwimp 12280*, FLAS); ZAF (*E.A. Schelpe 2115*, MO); ZAR (*L. Hauman 888*, JE).

21. Grimmia hamulosa Lesq., Mem. Calif. Acad. Sci. 1: 14. 1868. PROTOLOGUE: [U.S.A. California] Hab. gravelly soil, Mt. Dana, 10,000 ft., Bol. [H.N. Bolander]. TYPE: Gravelly soil. Mt. Dana, California (Bolander) (lectotype, designated by Muñoz (2000c), NY!; isolectotypes, FH! 2 replicates).

Grimmia brevirostris R.S. Williams, Bryologist 23: 52, pl. 3. 1920. PROTOLOGUE: [U.S.A.] Mountains near Buck's Valley, Plumas Co., Cal. [California], at 5600 ft. [sic], on granite. Collected by J.B. Leiberg, July 1900, no. 5445. TYPE: California [...] on granite rocks, mountains adjacent to Buck's valley, Plumas Co. Alt. 6500 ft. Collected by John B. Leiberg, 1900 July (lectotype, designated by Muñoz (2000c), NY!).
Synonymized by Muñoz (2000c).

DISTRIBUTION: USA (*A.T. Whittemore 3042*, MO).

22. Grimmia handelii Broth., Akad. Wiss. Wien Sitzungsber., Math.-Naturwiss. Kl., Abt. 1, 133(10): 567. 1924. PROTOLOGUE: [China] Prov. Setschwan austro-occid.: In montis Tschahungnyotscha trans flumen Yalung ad septentr. oppidi Yenyüen, 28°15', regione alpina, ad rupes schistosas; ca. 4150–4300 m (Nr. 2667). TYPE: Prov. Setschwan austro-occid.: In montis Tschanhungnyotscha supra vicum Ngaitschekou trans flumen Yalung ad septentr. oppidi Yenyüen ca. 28°15' siti regione alpina, ad rupes. Substr. schistaceo; alt. s.m. ca. 4150–4300 m. Leg. 27.V.1914 Dr. Heinr. Frh. v. Handel-Mazzetti (Diar. Nr. 498) (holotype, H-BR!; isotype, PC!).

DISTRIBUTION: CHN (*H.F.v. Handel-Mazzetti 2667*, H-BR).

23. Grimmia hartmanii Schimp., Syn. Musc. Eur.: 214. 1860. *Grimmia sphaerocarpa* Stirt. ex Braithw., Brit. Moss Fl. 2: 20. 1888, nom. inval., pro synon. *Dryptodon hartmanii* (Schimp.) Limpr., Laubm. Deutschl. 1: 789. 1889. *Grimmia sphaerophora* Stirt. ex Braithw., Brit. Moss Fl. 3: 263. 1905, nom. inval., orthogr. err. *Grimmia hartmanii* var. *septentrionalis* Loeske, Laubm. Eur. Part I: 144. 1913, nom. illeg., incl. typ. spec. *Grimmia hartmanii* subsp. *vulgaris* Loeske, Biblioth. Bot. 101: 177. 1930, nom. illeg., incl. typ. spec. *Grimmia hartmanii* subsp. *euhartmanii* Giacom., Atti Ist. Bot. Lab. Crittog. Univ. Pavia, ser. 5, 4: 223. 1947, nom. inval. *Grimmia hartmanii* var. *euhartmanii* Podp., Consp. Musc. Eur.: 289. 1954, nom. inval. PROTOLOGUE: ad saxa quarzosa umbrosa per Vogesum [France] sylvaticum praeprimis superiorem, etiam in Helvetia [Switzerland] prope Mornai ad latus meridionale m. Salève, ubi in molibus graniticis erraticis sub umbra Castaneae vescae copiose vidi; in Suecia [Sweden] copiosissime prope Holmiam atque Upsaliam provenit. TYPE: *Grimmia incurva* Schw. Circa Holmiam vulgaris, in Wermlandia rarus. 1854, leg. S.O. Lindberg (lectotype, designated by Geissler & Maier (1995: 501), BM!).

Grimmia lubtomeda Rabenh., Bryoth. Eur. fas. 11, nº 510. 1862, nom. nud. *Grimmia subtumida* Schimp. ex A. Jaeger, Ber. Tätigk. St. Gallischen Naturwiss. Ges. 1872–73: 75 (Gen. Sp. Musc. 1:

357). 1874, nom. inval., orthogr. err. ORIGINAL SPECIMENS: [Germany] auf Braunkohlensandstein in Siebengebirge bei Bonn. Januar 1861 leg. P. Dreesen (MO!, FH!, PC!).

Grimmia hartmanii f. *propagulifera* Milde, Bryol. Siles.: 150. 1869. *Dryptodon hartmanii* f. *propagulifer* (Milde) Limpr., Laubm. Deutschl. 1: 790. 1889 ["propagulifera"]. PROTOLOGUE: [Poland/Czech Republic border] Häufig im Riesengebirge. TYPE: Riesengebirge. Leg. Dr. Milde (lectotype, here designated, FH!; isolectotype, FH!).

Grimmia brotheri Lindb. ex Broth., Acta Soc. Sci. Fenn. 19(12): 89. 1892. *Dryptodon brotheri* (Broth.) Warnst., Hedwigia 53: 305. 1913. PROTOLOGUE: [Georgia] Cauc. major occ.: Radscha, ad saxa silicea regionis silvaticae mediae pr. Uzeri, cum fructibus deoperculatis, sat numerosis; Svania, in silvis jugi montani inter Enasch (Latal) et Betscho 15–1,600 m., cum fructibus deoperculatis (E.L. n. 192!). TYPE: Imeretia: ad saxa umbrosa pr. Utseri [Uzeri]. 7-VII-1877. V.F. Brotherus 301 (lectotype, here designated, H-SOL!).

Racomitrium patens var. *brachydictyon* Cardot, Bull. Herb. Boissier, sér. 2, 8: 333. 1908. *Grimmia patens* var. *brachydictyon* (Cardot) Iishiba, Nipponsan Senrui Sosetsu: 77. 1929. *Grimmia brachydictyon* (Cardot) Deguchi, J. Sci. Hiroshima Univ., Ser. B, Div. 2, Bot. 16: 173. 1979 ["1978"]. PROTOLOGUE: Japon: Namikawa, rochers, à 800 m (n. 2807). TYPE: Japon: Namikawa, rochers, à 800 m. Leg. Faurie, 1904. nº 2807 (holotype, PC!; isotypes, H-BR!, NY!).

Grimmia borneensis Dixon & Herzog, Hedwigia 66: 344. 1926. PROTOLOGUE: Borneo: Ohne nähere Fundortsangabe leg. Missionsinspektor R. Wegner, 1913. TYPE: Fl. v. Borneo. Leg P. Wegner 1913. Com. Kostrik (holotype, JE!).
Synonymized by Deguchi (1986: 328).

Grimmia hartmanii f. *atra* H. Möller, Ark. Bot. 26A(2): 91. 1934. PROTOLOGUE: [Sweden] Gästrikland. Torsåker, Korsån 1911 M[ihi: H. Möller]. TYPE: Gästrickland. Torsåker, Korsån. 15.8.1911 Leg. Hj. Möller (lectotype, here designated, S!).

DISTRIBUTION: ARM (*V. Manakyan 1802a*, ERE); AUT (*H. Winter s.n.*, JE); BEL (*F. Gravet s.n.*, PC); BGR (*R. Düll 1c*, JE); BIH (*J. Głowacki s.n.*, JE); CHE (*P. Culmann s.n.*, MA); CZE (*A. Geheeb s.n.*, FH); DEU (*L. Meinunger 1118*, JE); DNK (*C. Jensen s.n.*, FH); ESP (*P. Seró B-357*, BCB); FIN (*Brotherus s.n.*, JE); FRA (*J.-P. Frahm 935073*, MA); GBR (*D.G. Long 13469*, MA); GEO (*A.L. Abramova s.n.*, LE); HUN (*L. Vajda s.n.*, MA); ITA (*J.L. De Sloover 23980*, MA); JPN (*E. Iishiba 7740*, MO); NOR (*Janzen s.n.*, JE); POL (*J. Milde s.n.*, FH); ROM (*T. Herzog s.n.*, JE); SFE (*V. Ardeeva s.n.*, LE); SUN (*A.L. Abramova & I.I. Abramov s.n.*, LE); SVK (*H. Sedlmayr s.n.*, MO); SWE (*J.E. Zetterstedt s.n.*, H); TUR (*Davis & Hedge D.30403a*, MA); USA (*T.P. James s.n.*, MO); YUG (*D. Katić s.n.*, JE).

24. Grimmia humilis Mitt., J. Linn. Soc., Bot. 12: 100. 1869. PROTOLOGUE: [Chile] Magellan, ad Sandy Point, Lechler. TYPE: Sandy Point in pret. magell. Lechler (lectotype, designated by Muñoz (1999a: 140), NY; isolectotypes, BM!, PC!).

DISTRIBUTION: ARG (*A. Donat 110*, JE); CHL (*H. Deguchi 26027*, HIRO).

25. Grimmia incrassicapsulis B.G. Bell, Brit. Antarct. Surv. Bull. 63: 80. 1984. *Grimmia inaequalis* Dixon & Sainsbury, Trans. & Proc. Roy. Soc. New Zealand 75(2): 174. 1945, nom. illeg., non Stirt. 1908. PROTOLOGUE: [New Zealand] On rock, Mount Watkin, near Waikouaiti, Otago, ca. 1500 feet; coll. Simpson and Thomson, 30/11/1935, No. 823, type. Same locality and collectors, 13/7/1935, No. 811. TYPE: Mt. Watkin, Otago, N.Z. Simpson & Thomson n° 823. type (holotype, BM!; isotype, NY!).

DISTRIBUTION: NZL (*G. Simpson & Thomson 811*, BM); SGS (*B.G. Bell 3312*, BM).

26. Grimmia incurva Schwägr., Sp. Musc. Frond. Suppl. 1(1): 90. 1811. *Grimmia contorta* f. *incurva* (Schwägr.) Molendo, Ber. Naturhist. Vereins Passau 10: 119. 1875. *Grimmia incurvata* Schwägr. ex Grout, Bryologist 14: 75. 1911, nom. inval., orthogr. err. PROTOLOGUE: [Austria] Legi in summis alpibus Carinthiae, Glockner aliisque similibus primaevis, in Sudetum fossis nivalibus sterilem observavit C. Ludwig. TYPE: Glockner (lectotype, designated by Deguchi (1978: 190), G!; syntype ["In fossis nivalibus Sudetum"], B!).

Dicranum contortum Wahlenb., Fl. Carpat. Princ.: 346, tab. 4. 1814. *Campylopus contortus* (Wahlenb.) Brid., Muscol. Recent. Suppl. 4: 74. 1818 ["1819"]. *Dryptodon contortus* (Wahlenb.) Brid., Bryol. Univ. 1(1): 199. 1826. *Grimmia contorta* (Wahlenb.) Arn., Mém. Soc. Linn. Paris 5: 243. 1827. *Grimmia contorta* f. *typica* Chal., Pamietn. Fizyogr. 2: 40. 1882, nom. alt., nom. inval. *Grimmia contorta* f. *longifolia* Chal., Pamietn. Fizyogr. 2: 40. 1882, nom. alt., nom. illeg. *Grimmia incurva* f. *longifolia* (Chal.) Vilh., Věstn. Král. Ceské Společn. Nauk, Tř. Mat.-Přír. 1924: 31. 1924, nom. inval. *Grimmia incurva* f. *typica* (Chal.) Vilh., Věstn. Král. Ceské Společn. Nauk, Tř. Mat.-Přír. 1924: 31. 1924, nom. inval. PROTOLOGUE: [Sweden] in cryptis saxorum ad lacum alpinum Langesee im Völksgrund longe supra terminum Mughi copiose. TYPE: Ad Langensee vallis Völkensis. 10/10/1813 leg. G. Wahlenberg (isotype, H-SOL!).
Synonymized by Deguchi (1978: 190).

Grimmia contorta f. *tatrensis* Chal., Pamietn. Fizyogr. 2: 40. 1882, nom. alt. *Grimmia contorta* f. *brevifolia* Chal., Pamietn. Fizyogr. 2: 40. 1882, nom. alt. *Grimmia contorta* var. *tatrensis* (Chal.) Chal., Enum. Musc. Frond. Tatr.: 51. 1886. *Grimmia incurva* var. *tatrensis* (Chal.) Herzog, Bot. Jahrb. 47: 479. 1912. *Grimmia incurva* f. *brevifolia* (Chal.) Vilh., Věstn. Král. Ceské Společn. Nauk, Tř. Mat.-Přír. 1924: 32. 1924. *Grimmia incurva* f. *tatrensis* (Chal.) Vilh., Věstn. Král. Ceské Společn. Nauk, Tř. Mat.-Přír. 1924: 32. 1924. PROTOLOGUE: [Poland] Brzegi Morskiego Oka. Pyszna Hala w górze. Dolina Małego Kolbachu, wprost Łomnicy. Pod Wagą, nad Zmarzłym stawem. Kończysta szczyt i nad Batyżowiecką doliną, tudzież od strony Tupéj. Tomanowa, szczyt. Polski Grzebień od strony Felki i od Zmarzłego. Grań od Mięguszowskiego szczytu do Mnicha. Rysy szcsyt. Krywań, szcsyt i poniżéj. Lodowy, szcsyt i poniżéj od strony doliny Jaworzyńskiéj. Gierlach, szcsyt, tudzież poniżéj, od strony Batyżowieckiéj doliny. TYPE: Nad Zmarzłym nad Wagą. 6-VIII-79 [T. Chałubiński] (lectotype, designated by Ochyra & Cisło (1999: 79), ZAMU!).

Grimmia hagenii Kaurin, Nyt Mag. Naturvidensk. 31: 217. 1888 ["Hageni"]. *Grimmia incurva* var. *hagenii* (Kaurin) C.E.O. Jensen, Meddel. Grønland 15: 426. 1898. *Grimmia contorta* var. *hagenii* (Kaurin) Paris, Index Bryol. Suppl.: 172. 1900. PROTOLOGUE: [Norway] Planten er funden af Doktor Hagen paa Galdhø den 1ste August 1887 i omtrent 1800 Meters Høide o. H. paa Klipper (Gabbro?) [The plant was found by Dr. Hagen at Galdhø in 1-August-1887 at approximately 1800 m on cliffs (Gabbro?)]. TYPE: Lom: Galdhø 1800 mtr. 11 Aug. 1887. Leg. I. Hagen (lectotype, here designated, S!; isolectotypes, S! 2 replicates).

Coscinodon japonicus Sakurai, J. Jap. Bot. 19: 407, fig. 9. 1943. PROTOLOGUE: [Japan] Honsyu: Prov. Sinano, Sugadaira, Nekodake am Gipfel [...] (Leg. N. Takaki Typus in Herb. K. Sakurai Nr. 14505 Juli 1942; Nr. 877 von Takaki ist dasselbe). TYPE: (holotype, Deguchi (1978: 191), MAK not seen).
Synonymized by Deguchi (1978: 191).

Grimmia torngakiana Brassard & Hedd., Mem. New York Bot. Gard. 45: 216. 1987. PROTOLOGUE and TYPE: Canada, Labrador; Nachvak Fiord, Minaret Plateau, 58°55'N, 64°00'W, 1700 m elevation, turfs on mineral soil in seepage from late snowbeds, 22 Jul 1983, Hedderson 1893 (holotype, NFLD!).

DISTRIBUTION: AUT (*H.F.v. Handel-Mazzetti s.n.*, MA); CAN (*T.A. Hedderson 1892*, NFLD); CHE (*F. Winter s.n.*, MO); CZE (*Z. Pilous s.n.*, MO); DEU (*L. Loeske s.n.*, MO); ESP (*R.M. Cros s.n.*, BCB); FIN (*V.F. Brotherus s.n.*, JE); FRA (*W.P. Schimper s.n.*, H-SOL); GBR (*J.B. Duncan 5431*, MO); GEO (*A.H. Brotherus & V.F. Brotherus 128*, H-SOL); GRL (*W.S. Benninghoff & H.C. Robbins 7034*, NY); HUN (*A. Boros s.n.*, BP); ITA (*A. Giacomini s.n.*, JE); JPN (*Z. Iwatsuki s.n.*, W); NOR (*N.C. Kindberg s.n.*, S); POL (*H. Bednarek-Ochyra & R. Ochyra 220/90*, MO); PRT (*G. da Cunha & L. Sobrinho 13*, LISU); ROM (*A. Boros s.n.*, MA); SFE (*C. Wright 78*, FH); SIB (*E. Ignatova 6/18*, MA); SUN (*E.A. Ignatova 6/17*, MO); SVK (*L. Vajda s.n.*, BP); SWE (*G. Wahlenberg s.n.*, H-SOL); TWN (*Peng Ching-I 83-35A*, MO); TZA (*C. Uhlig s.n.*, H-BR); UKR (*A. Boros s.n.*, BP); USA (*G. Allen 1434*, MO).

27. Grimmia indica (Dixon & P. de la Varde) Goffinet & Greven, J. Bryol. 22: 141. 2000. *Trigonodictyon indicum* Dixon & P. de la Varde, Ann. Cryptog. Exot. 1: 40–42, fig. 3. 1928. PROTOLOGUE: [India] Pambar Torrent, Kodaikanal, Pulney Hills [sic]. févr. 1927. Leg. R. P. Foreau. TYPE: Pambar Torrent, Kodaikanal, Palni Hills, S. India. 8.2.1927. Collector: R. P. Foreau. In Herb. H. N. Dixon Ref. No. 815 (lectotype, here designated, BM!; isolectotypes, FH!, MO!, NY!).

Grimmia apophysata Gangulee, Nova Hedwigia 12(3/4): 428, pl. 87. 1967. *Grimmia apophysata* A. Jaeger, Ber. Tätigk. St. Gallischen Naturwiss. Ges. 1872-73: 66 (Gen. Sp. Musc. 1: 348). 1874, nom. nud. PROTOLOGUE: [India] Kurz 2344 lectus ad Phalut prope Darjeeling ad alt. 3,500 m. die 16.X.1868 et servatus in CAL. TYPE: 2344. Sikkim, Phallort 11,000'. S. Kurz (isotype, H-BR!).

DISTRIBUTION: IND (*C. Srinivasan 122*, FH).

28. Grimmia involucrata Cardot, Rev. Bryol. 36: 105. 1909. PROTOLOGUE: [Mexico] Etat de Hidalgo: Cuzamaloza [sic], rochers, en compagnie de *G. ovata* Web. et Mohr, 1908 (n. 10598 in parte). TYPE: Pringle, Plantae mexicanae nº 10598 p.p. On rocks, Cuyamaloya, 8500 ft. 17, Sept., 1908. Herb. Cardot (lectotype, designated by Muñoz (1999a: 141), PC!; isolectotype, PC!).
Synonymized with *G. affinis* Hornsch. by Crum (1994: 396).

NOTE: In the description of the species in Muñoz (1999a: 141) the capsule shape is erroneously reported as ventricose. It is symmetric, as it was correctly indicated in the key.

DISTRIBUTION: MEX (*C.G. Pringle 10599*, TENN).

29. Grimmia kidderi James, Bull. Torrey Bot. Club 6: 54. 1875. *Grimmia kilderi* James ex Paris, Index Bryol. ed. 2: 277. 1904, nom. inval., orthogr. err. PROTOLOGUE: Southern part of the Kerguelen Island, Lat. 49°21'S, Long. 70°15'E. TYPE: Kerguelen Island. Leg. Dr. J.H. Kidder 1874 (lectotype, designated by Muñoz (1999a: 143), FH!; isolectotypes, FH! 2 replicates).

Grimmia frondosa James, Bull. Torrey Bot. Club 6: 54. 1875. PROTOLOGUE: Southern part of the Kerguelen Island, Lat. 49°21'S, Long. 70°15'E. TYPE: Kerguelen Island. Dr. J.H. Kidder (lectotype, designated by Muñoz (1999a: 143), FH!; isolectotypes, FH! 2 replicates, NY!, PC!).
Synonymized by Muñoz (1999a: 143).

Grimmia minutifolia Müll. Hal., Bot. Jahrb. Syst. 5: 81. 1884. PROTOLOGUE: Kerguelen. TYPE: Kerguelen. Foundery branch. Dr. F. Naumann Dcbr. 1874 (lectotype, designated by Ochyra & Hertel (1990: 25), PC!; isolectotype, H-BR!).
Synonymized by Ochyra & Hertel (1990: 25).

Guembelia immersoleucophaea Müll. Hal., in Neumayer, Int. Polarforsch., Deut. Exped. 2: 317. 1890. *Grimmia immersoleucophaea* (Müll. Hal.) Kindb., Enum. Bryin. Exot., Suppl. 2: 107. 1893. PROTOLOGUE: [South Georgia] Austro-Georgia, ad rupes im Brockenthale raro, 23. Januario 1883; ad rupes des Köppenberges, 19 Majo 1883 cum fructibus calyptratis. TYPE: Brockenthale. Süd-Georgien 23/I 83 Will (lectotype, designated by Ochyra & Hertel (1990: 25), HBG not seen; syntype ["Köppenberges, 19.V.1883, *Will*"], PC!).
Synonymized by Ochyra & Hertel (1990: 25).

Grimmia austrofunalis Broth., Deutsche Südpolar-Expedition 1901–1903, 8: 87, tab. 7 fig. 4. 1906, nom. illeg., non Müll. Hal. 1898. *Grimmia kerguelensis* Cardot, Bull. Mus. Hist. Nat. (Paris) 22(6): 338. 1916, nom. nov. PROTOLOGUE: Kerguelen: Felswand an der Südmire (Werth). TYPE: Kerguelen, 18/2/1903. Leg. E. Werth (holotype, H-BR!; isotypes, JE!, NY!, PC!, S!).
Synonymized by Ochyra & Hertel (1990: 25).

Grimmia nordenskioldii Cardot, Bull. Herb. Boissier, sér. 2, 6: 7. 1906. PROTOLOGUE: [South Georgia] Cumberland bay: Bore valley, mt. Duse. Royal bay: Moltke harbour. TYPE: Géorgie du Sud: Royal bay. Leg. Skottsberg, 1902. nº 303 (lectotype, designated by Ochyra & Hertel (1990: 25), S not seen; isolectotype, S not seen; syntypes ["Royal Bay, Moltke Harbour"], H-BR!, NY!).
Synonymized by Ochyra & Hertel (1990: 25).

Grimmia bossierei Cardot & Thér., Bull. Mus. Hist. Nat. (Paris) 22(6): 338. 1916. PROTOLOGUE: [Kerguelen] (Bossière). TYPE: Kerguelen. Leg. R. Bossiere, 1913 (holotype, PC!; isotypes, H-BR!, NY!, PC!).
Synonymized by Ochyra & Hertel (1990: 26).

Grimmia stenobasis Dixon, in Christoph., Results Norw. Sci. Exped. Tristan da Cunha 1937-1938, 48: 23. 1960. PROTOLOGUE: Tristan da Cunha: On loose cinders, slopes of Peak, above Settlement,

1600m., No. 1751 p.p. TYPE: Tristan da Cunha. Above Settlement, slopes of Peak, loose cinders, 1600m., 15 March 1938. leg. E. Christophersen No. 1751a. det. H.N. Dixon (lectotype, here designated, FH!).

DISTRIBUTION: ARG (*C. Skottsberg 371*, H-BR); ATA (*J. Healy 65*, FH); ATF (*J.P. Hébrard s.n.*, PC); CHL (*H.L. Gunckel 1768*, PC); SGS (*B.G. Bell 1181*, BM); SHN (*E. Christofersen & Mejland 1780*, BM); ZAF (*B.J. Huntley 184*, BM).

30. Grimmia laevigata (Brid.) Brid., Bryol. Univ. 1(1): 183. 1826. "Bryum trichodes hirsutie canescens, capitulis subrotundis erectis, in pediculis brevissimis" Dill., Cat. Pl. Circa Gissam: 224. 1718, nom. inval. "Bryum orbiculare pulvinatum, hirsutie canescens, var. capsulis erectis praedita" Dill., Hist. Musc.: 395, tab. 50 fig. 65 D, E. 1741, nom. inval. *Campylopus laevigatus* Brid., Muscol. Recent. Suppl. 4: 76. 1818 ["1819"]. *Grimmia levigata* Brid. ex Loeske, Laubm. Eur. Part I: 78. 1913, nom. inval., orthogr. err. PROTOLOGUE: In Tyrole Italico, item circa Romam, Neapolin et in insula Ischiâ in rupibus apricis 1806 abunde legi. Röhling etiam e Palatinatu [Germany], Schleicher ex Helvetiae [Switzerland] miserunt. TYPE: Italia. Rome. Napoli. Ischia 1806 (lectotype, designated by Cao & Vitt (1986: 181), B!).

Grimmia campestris Burchell ex Hook., Musci Exot. 2: 129. 1819. *Dryptodon campestris* (Hook.) Brid., Bryol. Univ. 1: 794. 1827. PROTOLOGUE: [South Africa] In territorio coloniali Capitis Bonae Spei, in regione Roggeveld, inter Jakhals Fonteyn et Kuylenberg;- terrestris, in campis aridis, parce. D. Burchell. TYPE: Ex herb. Hook. An original specimen of *Grimmia campestris* Hook. Musc. Ex. t. 129. Burchells Cape Mosses (isotype, H-SOL!).

Grimmia leucophaea Grev., Mem. Wern. Nat. Hist. Soc. 4: 87, pl. 6. 1822. *Dryptodon leucophaeus* (Grev.) Brid., Bryol. Univ. 1: 773. 1827. PROTOLOGUE: [United Kingom] In rupibus prope Edinburgum ad locum King's Park dictum. TYPE: Rocks King's Park. Edinb. R.K. Greville (lectotype, designated by Muñoz (1999a: 144), BM!; isolectotypes, FH!, NY! 3 replicates, PC!).

Dryptodon brevifolius Brid., Bryol. Univ. 1: 775. 1827. PROTOLOGUE: [Switzerland] In Helvetia habitat ubi Schleicher in rupibus detexit. TYPE: *Dryptodon brevifolius* N. Schleich. Herb. Cand. (lectotype, designated by Muñoz (1999a: 144), B!).
Synonymized by Muñoz (1999a: 144).

Grimmia leiocarpa Taylor, London J. Bot. 5: 44. 1846. PROTOLOGUE: [Australia] Swan River, Mr. James Drummond. TYPE: Swan River. NH [Nova Hollandia] James Drummond 1843 (lectotype, designated by Muñoz (1999a: 144), FH!; isolectotypes, H-SOL!, NY! 2 replicates).
Synonymized by Muñoz (1999a: 144).

Grimmia subleucophaea Müll. Hal., Bot. Zeitung (Berlin) 16: 164. 1858. PROTOLOGUE: [Ethiopia] Abyssinia, Semen, ubi vulgatissimam legit Octobr. 1850 W. Schimper. Coll. II. No. 41 [*Gr. leucophaea* W.P. Schimp. in Musc. Abyss. Schimp.]. TYPE: 41. Auf felsblöcken von 9000 für 14000'. [illegible] 2 nov 50 9200' [illegible] (lectotype, designated by Ochyra in sched., BM!).

Grimmia crassa Müll. Hal. ex A. Jaeger, Ber. Tätigk. St. Gallischen Naturwiss. Ges. 1872–73: 82 (Gen. Sp. Musc. 1: 364). 1874, nom. nud. ORIGINAL MATERIAL: [South Africa] Africa austral., Tafelberg ad prom. bon. spei. ORIGINAL SPECIMEN: Tafelberg am Cap. (NY!).

Grimmia calotricha Müll. Hal., Linnaea 43: 454. 1882. PROTOLOGUE: Argentinia Cordobensis, Sierra de Cordoba, Las Peñas, Febr. 1871. [leg. P.G. Lorentz]. TYPE: Sierra de Cordoba, Las Peñas. 1871 (lectotype, designated by Muñoz (1999a: 144), BM!; isolectotype, PC!).
Synonymized by Muñoz (1999a: 144).

Grimmia leucophaea var. *latifolia* Limpr., Bryoth. Siles. n° 163. 1867, nom. nud. *Grimmia leucophaea* var. *latifolia* Limpr., Laubm. Deutschl. 1: 742. 1889. *Grimmia campestris* var. *latifolia* (Limpr.) Paris, Index Bryol. ed. 2: 269. 1904. *Grimmia campestris* f. *latifolia* (Limpr.) Mönk., Laubm. Eur.: 356. 1927. *Grimmia laevigata* f. *latifolia* (Limpr.) Podp., Consp. Musc. Eur.: 276. 1954. PROTOLOGUE and TYPE: [Poland] Spitzberg bei Hirschberg. Gesammelt Ostern 1867 [Limpricht, Bryoth. sil. No. 163 (1867)] (lectotype, designated by Muñoz (1999a: 144), FH!; isolectotype, JE!).
Synonymized by Muñoz (1999a: 144).

Grimmia sarcocalyx Kindb., in Macoun, Bull. Torrey Bot. Club 17: 271. 1890, nom. nud. *Grimmia sarcocalyx* Kindb., in Macoun, Cat. Canad. Pl., Musci: 66. 1892. *Grimmia campestris* subsp. *sarcocalyx* (Kindb.) Kindb., Eur. N. Amer. Bryin. 2: 219. 1898. PROTOLOGUE: [Canada] On rocks on the summit of the mountain between the Nicola and Thompson rivers at Spence's Bridge, B.C., alt. 3,500 feet, May 28th, 1889 (Macoun.). TYPE: N. Amer., Br. Columb., Spence's Bridge, 3000' 28/5 89. J. Macoun. C.M. [Carl Müller] vid. (lectotype, designated by Muñoz (1999a: 145), S!; isolectotype, CANM-198098!).
Synonymized by Muñoz (1999a: 145).

Grimmia bellii R. Br. bis, Trans. & Proc. New Zealand Inst. 27: 416, pl. 33 fig. 24. 1895 ["belli"]. PROTOLOGUE: [New Zealand] Dry rocks, Lyttelton hills, in several places; 1882. Birdling's Flat, Forsyth; Mount Self, Gorge of the Hurunui; 1893. Palmer Pass, head of the River Conway; February, 1894. Collected in these localities by R.B. TYPE: Lyttelton Hills Ch. Ch. Collected by Robert Brown (lectotype, here designated, BM!).

Grimmia murina Müll. Hal., Hedwigia 36: 105. 1897. PROTOLOGUE: Argentinia Patagonica, ad pedem Sierrae Ventanae in rupibus, cum Catenularia Ventanae associata: P.G. Lorentz legit 27. Feb. 1881. TYPE: Patagonia, Sierra Ventana. Lorentz 1881 (lectotype, designated by Muñoz (1999a: 145), JE!; isolectotypes, JE!, PC!).
Synonymized by Deguchi (1984: 21).

Grimmia leucophaea var. *longirostris* G. Roth, Eur. Laubm. 1: 564, tab. 51 fig. 9. 1903. *Grimmia laevigata* var. *longirostris* (G. Roth) Podp., Consp. Musc. Eur.: 276. 1954. PROTOLOGUE: [Germany] An Basaltfelsen auf dem Ramsberg bei Laubach am 2. April 1903 in einem reich fruchtenden Rasen von mir entdeckt. TYPE: Ramsberg bei Laubach (Hessen). 2/IV/1903. G. Roth (lectotype, here designated, S!).

Grimmia serripila Dusén ex Paris, Index Bryol. ed. 2: 287. 1904, nom. nud. ORIGINAL MATERIAL: Am. merid. Chile. Dusen M. Am. austr. n. 366. ORIGINAL SPECIMEN: Chile N° 366 P. Dusén (JE! [mixed with *G. trichophylla*], PC!).

Grimmia glauca Cardot, Rev. Bryol. 32: 17. 1905. PROTOLOGUE: [France] [...] 24 février 1903, dans les environs de Charleville. TYPE: (a specimen origin.) Ardennes: Charleville: rocher schistoux sec à Belair. Leg. J. Cardot, 1903 (lectotype, here designated, S!; isolectotype, BM!).

Grimmia senilis Shaw, Cape Monthly Mag. 17: 380. 1878. PROTOLOGUE: [South Africa] Collected by McLea on rocks, Graaff-Reinet, and by Shaw at Colesberg; by Bolus first gathered at Graaff-Reinet. TYPE: (PRE? not seen).
Synonymized by Magill (1981: 277).

DISTRIBUTION: AND (*C. Sérgio s.n.*, LISU); ARG (*C.C. Hosseus 1856*, FH); ARM (*V. Manakyan 4873a*, ERE); AUS (*A.C. Beauglehole 73091*, MEL); AUT (*H.C. Funck s.n.*, H); BEL (*C. Delogne s.n.*, FH); BRA (*D.M. Vital 9196*, FLAS); CAN (*J. Macoun s.n.*, CANM); CHE (*S.E. Bridel s.n.*, B); CHL (*E. Barros 6877*, FH); CHN (*D.-K. Li 16137*, ALTA); CPV (*K. Byström 180,* PC); CZE (*J. Podpěra s.n.*, H); DEU (*A.C.H. Braun s.n.*, H-SOL); DNK (*A. Hesselbo s.n.*, FH); DZA (*L. Trabut 2121*, PC); ESP (*J.L. Rupidera 45*, SALA); ETH (*E.F. Gilbert 279A*, NY); FRA (*F.G. Kneiff & C.P.W. Maercker s.n.*, FH); GBR (*R.K. Greville s.n.*, BM); GEO (*G. Woronow 219*, PC); GRC (*H.K. Haussknecht s.n.*, JE); HUN (*L. Vajda s.n.*, MA); IND (*R.R. Stewart 13725*, FH); IRN (*P. Goldblatt 822a*, MO); ISR (*M. Bizot 10040,* PC); ITA (*C.H. Binstead 174*, NY); KAZ (*S.G. Nesterova 521*, MO); KEN (*T. Pócs et al. 9213/E*, FH); KGZ (*R. Kamelin s.n.*, LE); LSO (*J. van Rooy 3287*, NY); MAR (*M.R. Crosby 4989*, NY); MEX (*A.J. Sharp et al. 6086*, TENN); MNG (*W. Schubert M273*, MO); NZL (*D.H. Vitt 8532*, MO); PAK (*B.L. Burtt B726*, MA); POL (*J. Milde s.n.*, FH); PRT (*C. Sérgio et al. 30.1c*, MA); REU (*M.R. Crosby 8403*, MA); SIB (*E. Ignatova 13/24*, MA); SUN (*A.L. Abramova & I.I. Abramov s.n.*, NY); SVK (*Z. Pilous s.n.*, FH); SWE (*S.O. Lindberg s.n.*, H); SYR (*H.F.v. Handel-Mazzetti 1658*, FH); TUN (*J.-P. Frahm 813442*, MA); TUR (*E. Nyholm & A.C. Crundwell 1060*, F); TZA (*T. Pócs et al. 88124/A*, FH); TZK (*P. Polyakov s.n.*, LE); UKR (*E.I. Vysotskaya s.n.*, H); URY (*E.J. Gibert 1243*, PC); USA (*A. Brant 3598*, MO); UZB (*Hudaikulov s.n.*, LE); ZAF (*D. Hardy 4218*, NY).

31. Grimmia leibergii Paris, Index Bryol. 2: 528. 1895, nom. nov. *Grimmia pachyphylla* Leiberg, Bull. Torrey Bot. Club 20: 113, fig. 144. 1893, nom. illeg., non Müll. Hal. 1885. PROTOLOGUE: [U.S.A.] Habitat granite, gneissoid and slate rocks throughout Kootenai Co., Idaho. J.B. Leiberg, No. 250, 1890–92. TYPE: Post Falls. Kootenai Co., Idaho. Coll. J.B. Leiberg 250. April 14th, 1890 (holotype, US not seen; isotypes, NY! 5 replicates, PC!).

DISTRIBUTION: USA (*A.T. Whittemore 6584*, MA).

32. Grimmia lisae De Not., Muscol. Ital. Spic.: 15. 1837. *Grimmia muehlenbeckii* var. *lisae* (De Not.) Bott., in Venturi & Bott., Atti Soc. Crittog. Ital. 3: 217. 1884. *Grimmia trichophylla* subsp. *lisae* (De Not.) Boulay, Musc. France 1: 378. 1884. *Dryptodon lisae* (De Not.) Loeske, Stud. Morph. Syst. Laubm.: 111. 1910. *Grimmia trichophylla* var. *lisae* (De Not.) Bott., in Luisier, Brotéria, Sér. Bot. 14: 32. 1916. PROTOLOGUE: [Italy] Ad saxa in collibus Taurinensibus vulgatissimam detexit D.us Lisa. TYPE: In collibus taurinensibus. Leg. Lisa 1836 (lectotype, designated by Muñoz (1999a: 145), RO!; isolectotypes, H-SOL!, NY!, RO!).

Grimmia trichophylla var. *brachycarpa* De Not., Syllab. Musc.: 250. 1838. PROTOLOGUE: [Italy. Sardinia]. TYPE: [Italy. Sardinia] Sta. Barbara pr. Cagliari. 3/3/835 (lectotype, here designated, RO!).

Grimmia ancistrodes Durieu & Mont., Ann. Sci. Nat. Bot., Sér. 3, 12: 318. 1849. PROTOLOGUE: [Algeria] circa lacum El Hout detexit cl. Durieu. TYPE: Algèrie. Durieu (lectotype, here designated, PC!; isolectotypes, PC!, PC-Besch!).

Grimmia sardoa Müll. Hal., Syn. Musc. Frond. 1: 786. 1849. *Grimmia trichophylla* var. *sardoa* (Müll. Hal.) Boulay, Musc. France 1: 379. 1884. *Grimmia muehlenbeckii* var. *sardoa* (Müll. Hal.) M. Fleisch., Malpighia 7: 329. 1893, nom. inval., pro synon. *Grimmia lisae* subsp. *sardoa* (Müll. Hal.) Kindb., Eur. N. Amer. Bryin. 2: 223. 1898. *Dryptodon sardous* (Müll. Hal.) Loeske, Stud. Morph. Syst. Laubm.: 111. 1910 ["sardoa"]. *Grimmia trichophylla* subsp. *sardoa* (Müll. Hal.) Bott., in Loeske, Laubm. Eur. Part I: 133. 1913. PROTOLOGUE: [Italy] Sardinia, ad rupes montium: De Notaris Martio 1835 legit. TYPE: ad rupes in montius Sardinia australis. a D. DeNotaris (lectotype, designated by Muñoz (1999a: 145), RO!).
Synonymized by Muñoz (1999a: 146).

Grimmia californica Sull., Expl. Railroad Mississippi Pacific, Descr. Moss. Liverw.: 187, pl. 4. 1857. PROTOLOGUE: [U.S.A.] Oakland, opposite San Francisco; also near Sonora. TYPE: Trees and dry places, mountains near Oakland, California. J.M. Bigelow, April 5, 1854. No. 2.b (lectotype, designated by Muñoz (1999a: 146), FH!).
Synonymized by Muñoz (1999a: 146).

Grimmia flexiseta Schimp. ex Jard., Bull. Soc. Linn. Normandie, sér. 2, 9: 263, 264. 1875, nom. nud. ORIGINAL MATERIAL: [U.S.A.] sur les rochers à Saucelito, dans la baie de San Francisco [...] ORIGINAL SPECIMEN: California. San Francisco (NY!, PC!).

Grimmia canariensis Schimp. ex Paris, Index Bryol. 2: 520. 1895, nom. nud. ORIGINAL MATERIAL: [Spain] Husn. M. Can. n° 181a. - Afr.: Ins. Canar. ORIGINAL SPECIMEN: T. Husnot, Pl. Canarienses.- 1866. N° 181[a] - Tenerife. Los Órganos. Juin (PC!).

Grimmia canadensis Kindb., Eur. N. Amer. Bryin. 2: 226. 1898. PROTOLOGUE: [Canada] Amer. R. Can. Vanc. isl.: Macoun. TYPE: Canada, Vancouver isl. 26/5 1893 J. Macoun 1894 (lectotype, designated by Muñoz (1999a: 147), S!).
Synonymized by Muñoz (1999a: 146).

Grimmia azorica Renauld & Cardot, Bull. Soc. Roy. Bot. Belgique 41(1): 58. 1905. *Grimmia trichophylla* subsp. *azorica* (Renauld & Cardot) Luisier, Brotéria, Sér. Bot. 17: 42. 1919. PROTOLOGUE: [Portugal] Açores: Graciosa (Brown, Blanchy); Santa-Maria (Trelease). TYPE: n° 1393c. Sta. Maria. Collected by Wm. Trelease 6/28, 1896 (lectotype, designated by Muñoz (1999a: 147), PC!; isolectotype, MO!; syntype ["*Brown s.n.*, Graciosa"], PC!).
Synonymized by Muñoz (1999a: 146).

Grimmia sardoa var. *robusta* M. Fleisch. & Warnst., Bryotheca meridionalis n° 342. 1910, nom. nud. *Grimmia trichophylla* f. *robusta* (M. Fleisch. & Warnst.) Podp., Consp. Musc. Eur.: 288. 1954, nom nud. ORIGINAL SPECIMEN: Spanien: Sierra de Cordoba, bei Cordoba 400–600 m an Felsen. April 1908. M. Fleischer. [Fleischer & Warnstorf, Bryotheca europaea meridionalis, n° 342 (1910)] (FH!).

Grimmia trichophylla f. *propagulifera* H. Winter, Hedwigia 55: 101. 1914, nom. illeg., non Limpr. 1889. PROTOLOGUE: [Spain] Guimar (Teneriffa), auf Felsen am deutschen Sanatorium, [...]

[März-Mai 1912]. TYPE: *Grimmia trichophylla* forma *depauperata*. Guimar, auf felsen dem Sanatorium. Marz 1912 leg. Dr. Winter (lectotype, designated by Muñoz (1999a: 147), JE!). Synonymized by Muñoz (1999a: 146).

DISTRIBUTION: BEL (*J. Cardot s.n.*, PC); CAN (*J. Macoun s.n.*, CANM); CPV (*A. Chevalier s.n.*, PC); CYP (*Unger s.n.*, JE); DZA (*J. Feldmann 2222*, MA); ESP (*C. Sérgio et al. 20.8*, BCB); FRA (*F. Renauld s.n.*, MO); GBR (*D.A. Jones 1394*, FH); GRC (*P. Haritonidou s.n.*, MO); ISR (*T. Kushnir 675*, PC); ITA (*A. Bottini s.n.*, FH); MEX (*E. Palmer 114a*, MO); PRT (*F. Carreiro 1017*, PC); SAU (*W. Frey & H. Kürschner 1538*, MA); TUN (*C.J. Pitard s.n.*, PC); TUR (*H. Kürschner & G. Parolly 6967*, MA); USA (*A.T. Whittemore 3530*, MO); YUG (*J. Baumgartner s.n.*, FH).

33. Grimmia longirostris Hook., Musci Exot. 1: 62. 1818. PROTOLOGUE: [Ecuador] Locis scopulosis, apricis, in declivitate montis Chimborazo, altit. 2380 hexapod. super porphyrite. (Regnum Quitense) Humboldt et Bonpland. TYPE: H. 2670. N. 76. M. de Humboldt (lectotype, designated by Deguchi (1984: 21), BM!; isolectotype, BM!).

Grimmia affinis Hornsch., Flora 2(2): 443. 1819, nom. illeg., non Hornsch. 1819. *Dryptodon ovatus* var. *affine* (Hornsch.) Hartm., Handb. Skand. Fl. ed. 3: 271. 1838. *Grimmia ovata* var. *affinis* (Hornsch.) Bruch & Schimp., Bryol. Europ. (fasc. 25–28) 3: 123, tab. 255ß. 1845, nom. illeg., incl. typ. var. prior. *Grimmia ovalis* var. *affinis* (Hornsch.) Broth., Acta Soc. Sci. Fenn. 19(12): 86. 1892. *Grimmia ovalis* f. *affinis* (Hornsch.) Mönk., Laubm. Eur.: 360. 1927. PROTOLOGUE: [Austria] Auf Felsenblöcken der Tauernkette und an diese angränzender Gebirge in Gesellschaft mit Gr. incurva Schwaegr., Gr. obliqua mihi und Gr. ovata und reift ihre Früchte im Aug. und Sept. TYPE: Windissmattreyer Tauern (lectotype, designated by Deguchi (1978: 158), B!). Synonymized by Muñoz (1998a: 353).

Grimmia obliquata Host, Syn. Pl.: 567. 1797, nom. inval. *Grimmia obliqua* Hornsch., Flora 2(1): 84. 1819. *Grimmia obliquata* Host ex Brid., Bryol. Univ. 1(1): 180. 1826, nom. inval., pro synon. *Grimmia ovata* var. *obliqua* (Hornsch.) Huebener, Muscol. Germ.: 183. 1833. *Dryptodon ovatus* var. *obliquus* (Hornsch.) Hartm., Handb. Skand. Fl. ed. 3: 271. 1838. *Grimmia ovalis* var. *obliqua* (Hornsch.) I. Hagen, Kongel. Norske Vidensk. Selsk. Skr. (Trondheim) 1909(5). 26. 1909. *Grimmia ovalis* f. *obliqua* (Hornsch.) Mönk., Laubm. Eur.: 360. 1927. *Grimmia ovata* f. *obliqua* (Hornsch.) Loeske, Biblioth. Bot. 101: 133. 1930. PROTOLOGUE: [Austria] Crypt. Cent. II. Dec. V. Auf Felsenblöcken in den Kärnthner u. Salzburger Alpen und auf dem Heiligenbl. Tauern. TYPE: In alpibus Salisburgensibus. Dr. Hornschuch (lectotype, designated by Muñoz (1998a: 353), BM!; syntypes ["ex alp. Carinth."], BM! 3 replicates; ["Heiligenbluter-Tauern"] JE!). Synonymized by Muñoz (1998a: 353).

Grimmia patens Hornsch., Flora 2: 84. 1819. *Grimmia ovata* var. *patens* (Hornsch.) Huebener, Muscol. Germ.: 182. 1833. *Dryptodon ovatus* var. *patens* (Hornsch.) Hartm., Handb. Skand. Fl. ed. 3: 271. 1838. PROTOLOGUE: [Germany] Ist von unserm Freunde, Herr Apotheker Funck, auf dem Fichtelberge gefunden und in vielen seiner Hefte als *G. ovata* ausgegeben worde.. ["bey Berneck" Funck, Cryptogamische Gewäsche des Fichtelgebirg's, fasc. 4, n° 86 (1805 "1804")]. TYPE: An Felsen bey Berneck. [Funck, Cryptogamische Gewäsche des Fichtelgebirg's, fasc. 4, n° 86 (1805 "1804")] (lectotype, here designated, GOET!; isolectotype, JE!).

Grimmia affinis var. *macrocarpa* Nees & Hornsch., Bryol. Germ. 2(1): 145, tab. XXI, 13c. 1827. *Campylopus ovalis* var. *hyrsutus* Myrin, Kongl. Vetensk. Acad. Handl. 1831: 258. 1832, nom. illeg., incl. typ. var. prior. *Grimmia ovata* var. *hirta* Myrin, Ärsberätt. Bot. Arbeten Upptäckter 1831: 333. 1832, nom. inval., orthogr. err. PROTOLOGUE: [Austria] wächst auf verwitterten Urgebirgsfelsen in der Gastein (Funck!). TYPE: (lectotype, here designated, fig. 13c in tab. XXI, Nees, C.G., Hornschuch, F. & Sturm, J., Bryol. Germ.). After an unsuccessful search for specimens referable to this taxon, the illustration in the original publication is proposed as lectotype of the name. It unambiguously represents *Grimmia longirostris* Hook.

Grimmia affinis var. *ramosissima* Nees & Hornsch., Bryol. Germ. 2(1): 144, tab. XXI fig. 13b. 1827. *Grimmia affinis* var. *elongata* Nees & Hornsch. ex Huebener, Muscol. Germ.: 709. 1833, nom. illeg., incl. typ. var. prior. PROTOLOGUE: [Austria] wächst auf den genannten höchsten Alpen um Heiligenblut, aber immer höher als Ó, z. B. am Fufse der Leiterköpfe auf der Pasterze, bei dem sogenannten Schaflerloch auf der Leiteralpe, dem Malnitzer-Tauern u.s.w. (Hsch.). TYPE: Heiligenblut b. Dartl. (lectotype, designated by Muñoz (1998a: 353), BM!).
Synonymized by Muñoz (1998a: 353).

Grimmia cylindrica Nees & Hornsch., Bryol. Germ. 2(1): 161, tab. XXIII fig. 18. 1827. *Grimmia ovata* var. *cylindrica* (Nees & Hornsch.) Huebener, Muscol. Germ.: 183. 1833. *Grimmia ovata* var. *submutica* De Not., Syllab. Musc.: 246. 1838, nom. illeg., incl. typ. var. prior. *Grimmia ovalis* var. *cylindrica* (Nees & Hornsch.) I. Hagen, Kongel. Norske Vidensk. Selsk. Skr. (Trondheim) 1909(5): 26. 1909. *Grimmia ovalis* f. *cylindrica* (Nees & Hornsch.) Mönk., Laubm. Eur.: 360. 1927. PROTOLOGUE: [Austria] Wächst auf kiesigen und sandigen Stellen im Anlaufthal bei Böckstein in der Gastein (Funck!). TYPE: (lectotype, here designated, fig. 18 in tab. XXIII, Nees, C.G., Hornschuch, F. & Sturm, J., Bryol. Germ.). After an unsuccessful search for specimens referable to this species, the illustration in the original publication is proposed as lectotype of the name. It unambiguously represents *Grimmia longirostris* Hook.

Grimmia sciuroides Nees & Hornsch., Bryol. Germ. 2(1): 155, tab. XXII fig. 17. 1827. *Grimmia ovata* var. *sciuroides* (Nees & Hornsch.) Huebener, Muscol. Germ.: 183. 1833. *Dryptodon sciuroides* (Nees & Hornsch.) Hartm., Handb. Skand. Fl. ed. 3: 271. 1838. PROTOLOGUE: [Austria] in der Gegend von Heiligenblut (Hsch. 1816), auf der Zirbnitzalpe im Möllthal (Laurer!), im Anlaufthal in der Gastein (Funck! Laurer!). TYPE: [Austria] Am Heiligenblut Baum (lectotype, here designated, GZU!).

Grimmia sciuroides f. *elongata* Nees & Hornsch., Bryol. Germ. 2(1): 159, tab. XXII fig. 17c.a. 1827. PROTOLOGUE: [Austria] [...] auf den Kärnthner-Alpen, ohne nähere Bezeichnung des Standortes (Laurer! Hoppe!). TYPE: Herb. Nees v Esenb. (lectotype, here designated, G!).

Grimmia sciuroides var. *foliosa* Nees & Hornsch., Bryol. Germ. 2(1): 158, tab. XXII fig. 17a. 1827. PROTOLOGUE: [Austria] in der Gastein (Hsch. 1816), bei Döllach (Funck! Laurer!), auf der Pasterze (Laurer! Hoppe!). TYPE: Herb. Nees v Esenb. (lectotype, here designated, G!).

Grimmia sciuroides var. *laxa* Nees & Hornsch., Bryol. Germ. 2(1): 159, tab. XXII fig. 17c.b, d. 1827. PROTOLOGUE: [Austria] im Anlauf- und Rötschachthal in der Gastein (Funk! Laurer!), bei dem Tauernhaus und am Geschlöfs in der Windischmattrey (Laurer! Funck!), auf dem Malnitzer-Tauern (Mielichhofer! Laurer!). TYPE: Herb. Nees v Esenb. (lectotype, here designated, G!).

Grimmia sciuroides var. *rigida* Nees & Hornsch., Bryol. Germ. 2(1): 158, tab. XXII fig. 17b. 1827. PROTOLOGUE: [Austria] auf den Alpen um Heiligenblut (Hoppe! Funck! Laurer! Hsch.). TYPE: In alpibus pr. Heiligenblut (lectotype, here designated, GZU!; isolectotype, G!).

Grimmia schimperi Bruch & Schimp. ex Müll. Hal., Syn. Musc. Frond. 1: 792. 1849. PROTOLOGUE: [Ethiopia] (in Musc. Abyss. No. 584 [err. pro 484]) [...] Abyssinia: W. Schimper. TYPE: Abyssinia [Schimper, W. 1842, Schimperi iter abyssinicum, Sectio II, nº 484] (lectotype, designated by Muñoz (1998a: 353), NY!; isolectotype, PC!).
Synonymized by Muñoz (1998a: 353).

Grimmia neilgherriensis Müll. Hal., Bot. Zeitung (Berlin) 11: 62. 1853. *Grimmia nilghiriensis* Müll. Hal. ex Mitt., J. Linn. Soc., Bot. Suppl. 1: 44. 1859, nom. inval., orthogr. err. *Grimmia neelgheeriensis* Müll. Hal. ex Kindb., Enum. Bryin. Exot.: 61. 1888, nom. inval., orthogr. err. PROTOLOGUE: [India] *Gr. ovata* Mont. Musc. Neilgher. No. 54? [in truncis arborum et rupibus circa Dodabett lectus (Montagne, C. 1842. Cryptogamae Nilgherienses seu Plantarum cellularium in montibus peninsulae indicae Neel-Gherries dictis à Cl. Perrottet collectarum Enumeratio. Ann. Sci. Nat. Bot. Sér. 2, 17: 243–256]. TYPE: in montibus Neilgheriensibus Indiae leg. Cl. Perrottet sub nº 1583 (lectotype, designated by Muñoz (1998a: 353), PC!; isolectotype, PC!).
Synonymized by Muñoz (1998a: 353).

Grimmia columbica De Not., Mem. Reale Accad. Sci. Torino, ser. 2, 18: 447, fig. 7. 1859. PROTOLOGUE: [Ecuador] a clarissimo Osculati ad flumen Napo in Columbia meridionalis. TYPE: [Ecuador] ad river Napo Columbia. Leg. Osculati. Herb. De Not. (isotype, H-SOL!).
Synonymized by Muñoz (1998a: 353).

Grimmia khasiana Mitt., J. Linn. Soc., Bot. Suppl. 1: 45. 1859. PROTOLOGUE: [India] Hab. In mont. Khasian. reg. temp., ad rupem Kollong, inter *Cladonias*, J.D: Hooker et T. Thompson! (No. 279). TYPE: Kullong rocks. nº 279 (holotype, NY!).

Grimmia peruviana Sull., U. S. Expl. Exped., Musc.: 8, tab. 5A. 1860. PROTOLOGUE: Hab. Andes of Peru, at an elevation of 15.000 feet. TYPE: U.S. Ex. Ex. Wilkes 1838–1842. Andes of Peru. at an elevation of 15.000 feet (lectotype, designated by Muñoz (1998a: 353), BM!; isolectotypes, BM!, FH!, NY!, PC!).
Synonymized by Muñoz (1998a: 353).

Guembelia bogotensis Hampe, Linnaea 32: 142. 1863. *Grimmia bogotensis* (Hampe) A. Jaeger, Ber. Tätigk. St. Gallischen Naturwiss. Ges. 1872–73: 70 (Gen. Sp. Musc. 1: 352). 1874. PROTOLOGUE: [Colombia] Bojota Los Laches ad rupes, 2800 metr., Jul. Nº 2011. TYPE: Nº 2011. Nova Granata. Bogota. Los Laches. alt. 2800 metr. 1859 Jul. Lindig 2011 (lectotype, designated by Muñoz (1998a: 353), PC!; isolectotype, NY!, PC!).
Synonymized by Churchill (1989: 109).

Grimmia hausmanniana De Not., Comment. Soc. Crittog. Ital. 2: 102. 1866. *Racomitrium hausmannianum* (De Not.) Molendo, Ber. Naturhist. Vereins Passau 10: 127. 1875. *Grimmia ovata* subsp. *hausmannii* Kindb., Eur. N. Amer. Bryin. 2: 227. 1898. *Grimmia ovalis* f. *epilosa*

Podp., Consp. Musc. Eur.: 278. 1954, nom. illeg., non Warnst. 1904. PROTOLOGUE: [Austria] Monte Rittnerhorn nel Tirolo meridionale, Hausmann. TYPE: Rittner Horn. Tyrol austr. Leg. Hausmann (isotype, BM!).
Synonymized by Muñoz (1998a: 353).

Grimmia tenuicaulis Hampe ex A. Jaeger, Ber. Tätigk. St. Gallischen Naturwiss. Ges. 1872–73: 72 (Gen. Sp. Musc. 1: 354). 1874, nom. nud. ORIGINAL MATERIAL: America austral., Chile. ORIGINAL SPECIMEN: Chile Andes. Herb. Hampe (FH!, NY!).

Grimmia ovata var. *praecox* A. Kern., Flora exsiccata austro-hungarica nº 316. 1881. PROTOLOGUE and TYPE: [Austria] Tirolia centralis. In saxis schist. ad Trins in valle Gschnitz; 1200 mt. s.m. (lectotype, designated by Muñoz (1998a: 353), BM!; isolectotypes, FH! 2 replicates, GOET!).
Synonymized by Muñoz (1998a: 353).

Grimmia integridens Müll. Hal., Linnaea 43: 460. 1882. PROTOLOGUE: Argentinia subtropica Tucumanensis, in alpinis "der Cienega" cum *Psilopilo gymnostomulo* forsan in altitudine 11,000 ped., 1873 [leg. P.G. Lorentz]. TYPE: Argentinien: Tucuman, in der Cienaga, ca. 11000'. 1873 leg. P.G. Lorentz (lectotype, designated by Muñoz (1998a: 353), PC!).
Synonymized by Muñoz (1998a: 353).

Grimmia leucophaeola Müll. Hal., Linnaea 43: 456. 1882. PROTOLOGUE: Argentinia Tucumanensis, in alpinis Sierrae de Aconquija prope Tafi, Aprili 1872. [leg. P.G. Lorentz]. TYPE: Argentina. Tafi. 1872 (lectotype, designated by Muñoz (1998a: 353), JE!; isolectotypes, BM!, JE!, NY!).
Synonymized by Muñoz (1998a: 353).

Grimmia raphidostega Müll. Hal., Linnaea 43: 459. 1882. PROTOLOGUE: Argentinia subtropica Saltensis, Nevado de Castillo, in alpinis cum Andreaeaceis, 1873. [leg. P.G. Lorentz]. TYPE: Flora argentiniae subtropicae. Salta, Nevado de Castillo, in alpinis. 1873 leg. P.G. Lorentz (lectotype, designated by Muñoz (1998a: 353), JE!).
Synonymized by Muñoz (1998a: 353).

Grimmia vernicosula Müll. Hal., Linnaea 43: 458. 1882. PROTOLOGUE: Argentinia subtropica, Rio seco [sic] prope Orán, 1873. [leg. P.G. Lorentz]. TYPE: Argentina. Oran. Lorentz (lectotype, designated by Muñoz (1998a: 353), JE!; isolectotype, BM!).
Synonymized by Muñoz (1998a: 353).

Grimmia rufa Müll. Hal., Bot. Jahrb. Syst. 5: 81. 1884. PROTOLOGUE: Kerguelen [Foundery branch, Nov. 1874, Naumann]. TYPE: Kerguelen, XI 1874. Naumann (lectotype, designated by Muñoz (1999a: 149), PC!; isolectotype, H-BR!).
Synonymized by Muñoz (1999a: 149).

Grimmia calyculata Müll. Hal., Flora 71: 414. 1888. PROTOLOGUE: [Tanzania] Monte Kilimandscharo, ad finem sylvestrem superiorem inter 3000–4000 m. TYPE: Kilimandscharo. Hb. Müller (lectotype, here designated, PC-Besch!; isolectotype, BM!).

Grimmia leucophaeoides Müll. Hal. ex Kindb., Enum. Bryin. Exot., Suppl. 1: 91. 1889, nom. nud. ORIGINAL SPECIMEN: Nevado de Castillo. Argentinia. Lorentz (JE!).

Grimmia cavifolia Lindb. & Arnell, Kongl. Svenska Vetenskapsakad. Handl. 23(10): 103. 1890. *Grimmia canifolia* Lindb. & Arnell ex Broth., Nat. Pflanzenfam. 3(1): 452. 1902, nom. inval., orthogr. err. PROTOLOGUE: [Russia] Fl. Jen.[Flumen Jenisei], T. mont.: Takmak bei Krasnojarsk. TYPE: Sibiria: Jenisei, Krasnojarsk, 56° n.l. på urberg. 11/6/1876. H. Wilh. Arnell (lectotype, designated by Muñoz (1998a: 353), H-SOL!; isolectotypes, JE!, PC!, UPS!).
Synonymized with *G. ovalis* by Ignatov & Cao (1994: 89), but see Muñoz (1998a: 353).

Grimmia immergens Müll. Hal., Flora 73: 483. 1890. PROTOLOGUE: [Tanzania] Africa or. trop., Kilima-Ndscharo, 3000–4800 m: Dr. Hans Meyer 1889. TYPE: Insched.- Kilimandscharo. hb. Müller (lectotype, here designated, PC-Besch!; isolectotype, BM!).

Grimmia subovata var. *laevipila* Schimp. ex Paris, Index Bryol. 2: 538. 1895, nom. nud. ORIGINAL MATERIAL: [Bolivia] Am. merid.: ibid. [Mandon Pl. Boliv.] n. 1635. ORIGINAL SPECIMEN: Amasuyos viciniis Achacache, Putchani. Alt. Reg. alpina 3950 m. Jul. 1856. G. Mandon - Plantae Andium Boliviensium nº 1635 (BM! 2 replicates, PC!).

Grimmia dimorphula Müll. Hal., Nuovo Giorn. Bot. Ital., n.s. 3: 108. 1896. PROTOLOGUE: Prope cacumen montis Iu-kia-po ad saxa, Jun. TYPE: China interior, provincia Schen-si sept. in saxis cacuminis montis Iu-kia-po, ster. 27 Jun. 1894 legit Rev. Jos. Giraldi, detrm. C. Müller sub nº 877 (lectotype, designated by Cao & Vitt (1986: 170), H; isolectotypes, PC! 4 replicates).
Synonymized with *G. ovalis* by Cao & Vitt (1986: 170).

Grimmia nanoglobosa Müll. Hal. ex E. Britton, Bull. Torrey Bot. Club 23: 477. 1896. PROTOLOGUE: [Bolivia] Mapiri, 5000 ft. (3195). TYPE: Mapiri, Bolivia, 5,000 ft., May, 1886. Leg. H.H. Rusby, M.D. (holotype, NY!).
Synonymized by Deguchi (1987: 25).

Grimmia ortholoma Kindb., Rev. Bryol. 23: 19. 1896. PROTOLOGUE: Amer. Canada, Brit. Columbia: Macoun. TYPE: N. Amer., B.C., Dear Park, rocks. 8/6 90. J. Macoun (holotype, S!; isotype, S!).
Synonymized by Allen (1995: 161).

Grimmia breviexserta Müll. Hal., Bull. Herb. Boissier 5(3): 200. 1897. *Grimmia brevipes* Müll. Hal. ex Paris, Index Bryol. Suppl.: 172. 1900, nom. inval. PROTOLOGUE: [Guatemala] Quezaltenango, Coll. Nº 115, cum *Grimmia bernoullii* associata [leg. Bernoulli & Cario]. TYPE: Ex Museo Museo Berolinensi. Guatemala: Quezaltenango. Leg. Bernoulli et Cario (lectotype, designated by Muñoz (1998a: 353), PC!).
Synonymized by Muñoz (1998a: 353).

Grimmia micromitria Schimp. ex Müll. Hal., Nuovo Giorn. Bot. Ital., n.s. 4: 165. 1897, nom. nud. ORIGINAL MATERIAL: [Bolivia, Mandon] nº 1632. ORIGINAL SPECIMEN: G. Mandon. Plantae Andium Boliviensium nº 1632. Prov. Larecaja. Viciniis Sorata, Loma de Ullontiji, in scopulosis. Reg. temp. 2800 m. Feb. 1868 (BM!).

Grimmia micro-ovata Müll. Hal., Nuovo Giorn. Bot. Ital., n.s. 4: 128. 1897. PROTOLOGUE: [Bolivia] Germain [...] 1889 prope Choquecamata provinciae Cochabamba Bolivianae alto-montosae ex altitudine pedum 10–12,000, ut opinor, collectas. TYPE: Bolivia, provincia

Cochabamba prope Choquecamata. Jun. 1889 legit Germain determ. Dr. C. Müller sub nº 1142 (lectotype, designated by Muñoz (1998a: 353–354), JE!; isolectotypes, NY!, PC!).
Synonymized by Deguchi (1987: 25).

Grimmia semipilosa Hampe, in Müll. Hal., Linnaea 43: 461. 1882, nom. inval., pro synon. *Grimmia subovata* Schimp. ex Müll. Hal., Linnaea 43: 461. 1882, nom. inval., pro synon. *Grimmia subovata* Schimp. ex Müll. Hal., Nuovo Giorn. Bot. Ital., n.s. 4(2): 128. 1897. PROTOLOGUE: Bolivia, prov. Larecaja, Sorata, Ancouma, supra muros de la Hacienda Peñas, reg. alpina, 3700–4000 m. alta, Aprili 1860: G. Mandon. TYPE: Prov. Larecaja, vicinitis Sorata; Ancouma, supra muros de la hacienda Peñas. Regio alpina y subalpina, 3700–4000 m. April 1860. G. Mandon - Plantae Andium Boliviensium nº 1634 (lectotype, designated by Muñoz (1998a: 354), BM!; isolectotypes, BM! 2 replicates, FH!, G!, NY!, PC!).
Synonymized by Muñoz (1998a: 354).

Grimmia arctophila subsp. *labradorica* Kindb., Eur. N. Amer. Bryin. 2: 221. 1898. PROTOLOGUE: Amer. r. Can.: Northern Labrador: Macoun 1896. TYPE: Canada, N. Labrador 21/7 1896 leg. J. Macoun. Det. N.C. Kindberg (lectotype, designated by Muñoz (1998a: 354), S!; isolectotype, CANM-198076!).
Synonymized by Muñoz (1998a: 354).

Grimmia itatiaiensis Broth. ex Müll. Hal., Bull. Herb. Boissier 6(2): 108. 1898. PROTOLOGUE: Brasilia, Serra Itatiaia, ad rupes, 2100 m. alta. Martio 1894: E. Ule, Coll. Nº 1913. TYPE: Serra do Itatiaia, 2100 m., ad rupes, m. Martii 1894, leg. E. Ule. Ule & Brotherus, Bryotheca brasiliensis, nº 124 (lectotype, designated by Muñoz (1998a: 354), H-BR!; isolectotypes, GOET!, JE!, NY!, PC! 3 replicates, UPS!).

Synonymized with *G. affinis* by Cao & Churchill (1995: 505). These authors reported page 51 for the place of publication, but this is the page of the independently paginated reprint. The actual page of the journal is 108.

Grimmia itatiaiae Müll. Hal., Bull. Herb. Boissier 6(2): 109. 1898. PROTOLOGUE: Brasilia, Serra Itatiaia, 2300 m. alta, ad rupes der Agulhas Negras, Martio 1894, c. fr. immaturis paucis: E. Ule, Coll. Nº 1830. TYPE: Brasilia, Serra do Itatiaia, an Felsen der Agulhas-Negras. 2300 m, 3/1894. Leg. E. Ule 1830 (lectotype, designated by Muñoz (1998a: 354), H-BR!).
Synonymized by Muñoz (1998a: 354).

Grimmia commutata f. *brevipila* Broth., Musci turkestanici, nº 49. 1899, nom. nud. in sched. ORIGINAL SPECIMEN: [Kazakstan] Kungei Alatau: Kokoirok in valle fl. Koissu, ad saxa reg. alp., Jul. 13, 1896. leg. V.F. Brotherus (JE!).

Grimmia preussii Broth. ex Paris, Index Bryol. Suppl.: 175. 1900, nom. nud. ORIGINAL SPECIMEN: [Cameroon] Afr. occ., Kamerun-pii, auf Lava, 3500 m., 26 Febr. 1891. Leg. Preuss (H-BR!, PC!).

Grimmia elata Kindb., Rev. Bryol. 32: 33. 1905. PROTOLOGUE: 1. Found by J. Macoun in the Canadian Yukon-district 1902. TYPE: Canada, Yukon distr., Hunker creek 25/7 1902. Macoun (lectotype, designated by Muñoz (1998a: 354), S!; isolectotypes, CANM-198084!, H-BR!).
Synonymized by Muñoz (1998a: 354).

Grimmia ovataeformis Kindb., Rev. Bryol. 32: 35. 1905. PROTOLOGUE: 2. Found by J. Macoun 1904 in British Columbia, 5200–7500 feet above the sea. TYPE: Canada, B.C., lake Louise, rocks, 6000 f., 13/9 04. Macoun (holotype, S!; isotype, CANM-198092!).
Synonymized with *G. ovalis* by Allen (1995: 162), but see Muñoz (1998a: 353).

Grimmia praetermissa Cardot, Rev. Bryol. 36(5): 105. 1909. PROTOLOGUE: [Mexico] Jalapa (Bonpland, in herb. Museum Par.); Orizaba (Galeotti, n. 6974 in parte, in herb. Hort. Bot. bruxell.). Etat de Mexico: volcan de Toluca, 1892 (n. 26 a). TYPE: State of Mexico. In the crater of the Volcano of Toluca, 13,500 ft. 25.Sept.1892. Pringle, Plantae mexicanae nº 26a (lectotype, designated by Muñoz (1998a: 354), NY!; isolectotypes, FH!, JE!, PC! 4 replicates; syntypes ["*Galeotti 6974*"], PC!, TENN!).
Synonymized by Muñoz (1998a: 354).

Grimmia herzogii Broth., in Herzog, Biblioth. Bot. 87: 55. 1916. PROTOLOGUE: [Bolivia] Oberes Chocaya-Tal, alt. 4400 m, No. 3579; Yanakaka, alt. 4500 m, No. 3826. TYPE: An Felsen eines Gipfel der Yanakakabastion. ca. 4500 m. Leg. Th. Herzog Nº 3826. Juli 1911 (lectotype, designated by Deguchi (1987: 23), JE!; isolectotypes, H-BR!, PC!; syntypes ["*Herzog 3579*"], H-BR!, JE!, PC!).
Synonymized by Muñoz (1998a: 354).

Grimmia nigella Herzog, Biblioth. Bot. 87: 55, fig. 17. 1916. PROTOLOGUE: [Bolivia] An Felsen im oberen Chocayatal, ca. 4400 m, No. 3597 u. 3587/a; an Felsen bei der Saittulaguna, ca. 4300 m, No. 2679. TYPE: Plantae in itinere secundo per Boliviam lectae. An Felsen bei der Saittulaguna. ca. 4300m. Leg. Th. Herzog nº 2679. Juni 1911 (lectotype, designated by Muñoz (1998a: 354), JE!; syntypes ["*Herzog 3597*," "*Herzog 3587/a*"], JE!, ["*Herzog 3597*"]PC!).
Synonymized by Muñoz (1998a: 354).

Grimmia speirophylla Herzog, Biblioth. Bot. 87: 55. 1916. PROTOLOGUE: [Bolivia] An Felsen im Hochtal von Viloco, ca. 4600 m, No. 3148; an Felsen der Yanakakabastion, ca. 4500 m, No. 3745 u. 3827; an Felsen des Cerro Tunari, ca. 5000 m, No. 4871 u. 4811. TYPE: Plantae in itinere secundo per Boliviam lectae. An Felsen der Yanakakabastion. ca. 4500m. leg. Th. Herzog nº 3827. Juni 1911 (lectotype, designated by Muñoz (1998a: 354), JE!; isolectotype, JE!; syntypes ["*Herzog 4871*," "*Herzog 3148*"] JE!, ["*Herzog 4811*"] NY!, ["*Herzog 4811*"] PC!).
Synonymized by Muñoz (1998a: 354).

Grimmia speirophylla f. *humilis* Herzog, Biblioth. Bot. 87: 55. 1916. PROTOLOGUE: [Bolivia] [...] aus den Yanakakabergen (sine No.). TYPE: Flora von Bolivia. Torreni-Yanakaka. leg. Th. Herzog. Juli, 1911 (holotype, JE!).
Synonymized by Muñoz (1998a: 354).

Grimmia allionii Broth., Rev. Bryol. 47: 9. 1920. PROTOLOGUE: Prov. Azuay. Canar, ad rupes; 3.400 m. TYPE: America meridionalis, Ecuador SW. In rupibus montis prope Cañar, 3400 m, Azuay. 16 n. 1909. Legit Rev. M. Allioni (lectotype, designated by Muñoz (1998a: 354), H-BR!; isolectotype, H-BR!, PC!).
Synonymized by Muñoz (1998a: 354).

Grimmia afro-ovata Broth. & Thér., Bull. Mus. Hist. Nat. (Paris) 30: 240. 1924. PROTOLOGUE: Kenya (Mission Gromier-Le Petit, 1911). TYPE: Kenya (Kilimanjaro). Leg. Alluaud. (lectotype, designated by Muñoz (1998a: 354), PC!; isolectotype, H-BR!, PC!).
Synonymized by Muñoz (1998a: 354).

Grimmia catalinensis E.B. Bartram, Bryologist 27: 62, pl. 10. 1924. PROTOLOGUE: [U.S.A.] Rock faces in ravine near top of Mt. Lemon, Santa Catalina Mts., Pima Co., Arizona, 9000 ft. elevation. E.B. Bartram, No. 387, January 15, 1923. TYPE: Mosses of Southern Arizona. Rocks. Ravine near top of Mt. Lemon, Santa Catalina Mountains, Pima County. Altitude 9000 ft. Jan. 15 1923. Collected by Edwin B. Bartram (holotype, FH!; isotypes, FH!, H-BR!, NY!, MO!, PC!).
Synonymized with *Grimmia ovalis* by Jones (1933: 33), but see Muñoz (1998a: 354).

Grimmia catalinensis var. *mutica* E.B. Bartram, Bryologist 27: 62. 1924. *Grimmia ovalis* f. *mutica* (E.B. Bartram) G.N. Jones, Moss Fl. N. Amer. 2(1): 34. 1933. PROTOLOGUE: [U.S.A.] White House Canyon, Santa Rita Mts., Arizona, February 18, 1923, elevation 6000 ft., No. 683B. TYPE: Mosses of Southern Arizona. Low boulder. White House Canyon, Santa Rita Mountains, Santa Cruz County. Altitude 5500 ft. Feb. 18 1923. Collected by Edwin B. Bartram (holotype, FH!).
Synonymized by Muñoz (1998a: 354).

Grimmia somervellii Dixon, J. Bot. 63: 221. 1925. PROTOLOGUE: [Nepal] Rongbuk Valley, 16,000–17,000 ft. (1001 [holotype]); 16,000–17,500 ft. (1004, 1005); on soil, 16,500–17,500 ft., 15 miles N. of Base Camp, near Lamna La Pass (1006) [Somervell]. TYPE: Flora of Mt. Everest. 16000–17500 ft. Collector J.H. Somervell. Date 1924 (paratype, BM!).

Grimmia longipes Broth., Kryptogamae exsiccatae editae a Museo Palatino Vindobonensi nº 2980. 1926, nom. nud. in sched. ORIGINAL SPECIMEN: China (prov. Yünnan bor.-occid.): in rupibus diabasicis regionis temperatae prope vicum Ngulukö ad urbem Lidjiang ("Likiang"), ca. 3000 m s.m., m. Jun. c.fr. vet., m. Oct. c.fr. mat. leg. H. Handel-Mazzetti. specim. originale (F!, H-BR!, MA-7094!, PC!).

Grimmia ovalis f. *gracilis* H. Möller, Ark. Bot. 26A(2): 39. 1934. PROTOLOGUE: [Sweden] Västerbotten. Umeå, Esmarksberget 1896 C. P. Laestadius. [...] Lycksele lappmark. Stensele. Sandvik 1924 Sm. TYPE: Västerbotten. Umeå, Esmarksberget. 13.9.1896 C. P. Laestadius (lectotype, here designated, S!; syntype ["Lycksele lappmark. Stensele, Sandvik. 2/7 1924. *Carl Stenholm*"], S!).

Grimmia trollii Herzog, Hedwigia 74: 102. 1934. PROTOLOGUE: Bolivia: Curahuara (n. 58), Gipfel des Cerro de la Esperanza, Cordillera Real, 4500 m (n. 72) [C. Troll]. TYPE: Fl. v. Bolivia. Curahuara. leg. K. Troll nº 58 (lectotype, designated by Deguchi (1987: 23), JE!).
Synonymized by Deguchi (1987: 23) with *G. herzogii*, but see Muñoz (1998a: 354).

Grimmia cinerea Thér., Rev. Bryol. Lichénol. 9: 9. 1936. PROTOLOGUE: [Ecuador, leg. R. Benoist] Condorguachana, rochers, 4.150 m (nº 3153). TYPE: 3153. Equateur, rochers du Condorguachana. R. Benoist 4.XI.1930 (holotype, PC!; isotypes, PC! 2 replicates).
Synonymized by Churchill (1994: 2).

Grimmia stenopyxis Thér., Rev. Bryol. Lichénol. 9: 8. 1936. PROTOLOGUE: [Ecuador, leg. R. Benoist] Base du Pichincha, sur rochers d'andésite, 3.200 m (nº 3076, 2276); Rumipamba, à terre

dans les broussailles, 3.000 m (nº 2957); Environs de Quito, 2.800 m (nº 2281). TYPE: Equateur: base du Pichincha. rocher. R. Benoist 3076. 24.X.1930 (lectotype, designated by Muñoz (1998a: 354), PC!; isolectotype, F!; syntypes [“*Benoist 2281*”] F!, [“*Benoist 2957*”] NY!).
Synonymized with *G. affinis* by Cao & Churchill (1995: 507).

Grimmia sumatrana Dixon, Ann. Bryol. 12: 50. 1939. PROTOLOGUE: [Indonesia] G. Losir, on rock, 3250–3500 m, 4 Feb. 1937; coll. Van Steenis (10152); Hb. Bog (4035) [Buitenzorg Botanical Garden]. TYPE: (syntype, Deguchi (1986: 328–329), L not seen).
Synonymized by Deguchi (1986: 328–329).

Grimmia antillarum Thér., Rev. Bryol. Lichénol. 13(Trav. Husnot, 2): 13. 1944. PROTOLOGUE: [Dominican Republic] Santo-Domingo: Cordillera Central, prov. de Azua, Los Vallecitos de Yaque, 2.500 m (nº 13.630) [leg. M.E.L. Ekman, 1929–1930]. TYPE: Plantae Indiae Occidentalis nº 13630. Hispaniola. Civ. Santo Domingo: Cordillera Central, prov. de Azua, Los Vallecitos de Yaque, c. 2500 m. 2.X.1929. leg. E.L. Ekman. det. Thér. (lectotype, designated by Muñoz (1998a: 354), PC!; isolectotypes, F!, NY!, PC!).
Synonymized by Muñoz (1998a: 354).

Grimmia akaisialpina Takaki, Bot. Mag. (Tokyo) 64: 180, fig. 4. 1951. PROTOLOGUE: [Japan] South Alps; Sensui pass (2700 m, no. 10137, Aug. 15, 1950). TYPE: (holotype, Deguchi (1978: 158), private herbarium of Takaki, not seen).
Synonymized by Takaki et al. (1970: 186–187).

Grimmia ovata f. *glacialis* P. de la Varde, Ark. Bot., n.s. 3: 147. 1955, nom. nud. ORIGINAL MATERIAL: Teleki Valley, on frost-lifting ground, 4200 m (1711: forma glacialis); Nairobi Valley, 4250 m on frost-lifting ground (1909: f. glacialis). ORIGINAL SPECIMEN: Kenya Colony: Mt. Kenya, Teleki Valley, alpine regime on frost-lifting ground in the upper part of the valley. 4200 m.s.m. 27.7.1948 Coll. Olov Hedberg 1711. Det. PV 11909 (PC!).

Grimmia meridensis Vareschi, Flora de los Páramos de Venezuela: 87. 1970, nom. nud. ORIGINAL MATERIAL: [Venezuela] Andes. [...] cima del Páramo de las Piedras Blancas en una altitud de 4.762 m.

Grimmia maido Greven, Bryologist 99: 428, fig. 1. 1996. PROTOLOGUE: Africa. La Réunion. View-point le Maïdo. at the end of RF 8, along trail to Grand Bénare, alt. 2150 m, 1995. 10.07, H. C. Greven & S. N. Khoebal 4000.1. TYPE: Ile de la Réunion. Le Maïdo, at the end of R.F. 8, alt. 2150 m. Leg. H.C. Greven & S.N. Khoeblal 4000/1. 7 October 1995 (holotype, L not seen; isotypes, MO!; paratype, MO! [“*M.R. Crosby & C.A. Crosby 8430*”]).
Synonymized by Muñoz (1998a: 354).

DISTRIBUTION: ARG (*I.M. Lamb 5288*, FH); AUS (*C. Beauglehole 14712*, MEL); AUT (*C.F. Hornschuch s.n.*, B); BOL (*M. Lewis 79-26*, MO); BRA (*D.M. Vital 1708*, FLAS); BTN (*W. Griffith 127*, MO); CAN (*G.R. Brassard 15284*, FH); CHE (*P. Culmann s.n.*, MA); CHL (*R. Santesson 786*, FH); CHN (*B.C. Tan 93-1184*, FH); CMR (*J.P.M. Brenan in P.W. Richards 4269*, MO); COL (*A.M. Cleef 1798*, COL); CRI (*M. Crosby 3917*, TENN); CZE (*J. Duda s.n.*, BP); DEU (*J.C. Zenker & F.D. Dietrich s.n.*, FH); DOM (*A. & P. Liogier 25032*, FH); ECU (*B. Løjtnant 15345*, MO); ESP (*J.L. Rupidera 1*, SALA); ETH (*D.A. Petelin 37-7*, MO); FIN (*E.F. Lackström & V.F. Brotherus s.n.*, JE); FRA (*J.-P. Frahm 935069*, MA); GBR (*F.B. White s.n.*, NY); GEO (*C.C. Townsend 85/860*, ALTA);

GRL (*W.C. Steere 62-332*, NY); GTM (*A.J. Sharp 4771*, TENN); HND (*B. Allen 12260*, MO); IND (*J. Doulea 3110*, H-BR); ISL (*B. Jóhannsson 12769*, ICEL); ITA (*G. Venturi s.n.*, FH); JPN (*M. Mizutani 13651*, MO); KAZ (*B. Allen 10849*, MA); KEN (*C. Troll 5570*, JE); LKA (*C. Ruinard & A.H.M. Jayasuriy 17/40*, MO); LSO (*R.E. Magill 4186*, MO); MDG (*G. Cremers 1705*, MO); MEX (*Á. Cárdenas S. 5344*, MO); MNG (*W. Schubert M183*, MO); MWI (*N.G. Hodgetts M2123A*, MA); MYS (*B. Tan 89-521*, NY); NOR (*D.H. Norris 69524*, MO); NPL (*A. Zimmermann s.n.*, G); NZL (*J. Child M573*, F); PAK (*J.F. Duthie s.n.*, H-BR); PER (*E. Hegewald 5459*, MO); PHL (*B.C. Tan 82-195*, FH); PNG (*A. Eddy 2519*, BM); POL (*G.F. Kaulfuss s.n.*, BM); PRT (*H. Persson s.n.*, NY); REU (*J.L. De Sloover 17934*, MO); SFE (*B.C. Tan 97-492*, FH); SIB (*E. Ignatova 13/110*, MA); SUN (*I.V. Chernyadjeva 53*, LE); SVK (*Z. Pilous s.n.*, BP); SWE (*R.W. Hartman s.n.*, H); TKM (*G.J. Potanin s.n.*, LE); TWN (*M.-J. Lai 6886*, MO); TZA (*R.E. Gereau 1649*, MO); UGA (*G.H.S. Wood 1107*, PC); USA (*F.A. MacFadden 8036*, MO); VEN (*D. Griffin III et al. 2115*, FLAS); ZAF (*R.E. Magill 6607*, MO).

34. Grimmia macroperichaetialis Greven, Bryologist 101: 100–102. 1998. PROTOLOGUE: Australia. Victoria. Mt. Buller, summit, 1790 m, 14 March 1997, Greven Aus. 35 [paratypes listed]. TYPE: Australia. Victoria. Mt. Buller, summit, 1790 m, 14 March 1997, Greven Aus. 35 (holotype, CANB not seen; isotype, MO!).

DISTRIBUTION: AUS (*C. Beauglehole 15833*, MEL).

35. Grimmia macrotheca Mitt., J. Linn. Soc., Bot. Suppl. 1: 44. 1859. PROTOLOGUE: [India] In Himalayae orient. reg. temp., Sikkim, J.D. Hooker! (No. 316). TYPE: 316. Jongui. alt. 12.000. Dr. Hooker (holotype, NY!; isotypes, BM! 2 replicates).

Grimmia longicapsula C.H. Gao & T. Cao, Acta Bot. Yunnan. 3: 395, tab. 5 figs. 1–8. 1981. PROTOLOGUE: [Tibet] Xizang: Districtus Yadong Xian, ad rupes, 28.V.1975, Zang Mu 39 (Typus in ISPH, Isotypus in HKAS). TYPE: Tibet. Yadong County. Ji-Neou. Zang Mu 28 V '75, n° 39 (holotype, IFSBH!; isotype, NY!).
Synonymized by Cao & Vitt (1986: 196).

DISTRIBUTION: CHN (*S. He 31583*, MO).

36. Grimmia mammosa C.H. Gao & T. Cao, Acta Bot. Yunnan. 3: 394, tab. 4 figs. 1–9. 1981. *Grimmia mammillaria* X.J. Li & M. Zang, Bryoflora of Xizang: 138. 1985, nom. inval., orthogr. err. PROTOLOGUE: [Tibet] Xizang: Districtus Yadong Xian, ad rupes, alt.s.m. 2900 m, 29.V.1975, Zang Mu 71 (Typus in ISPH, Isotypus in HKAS). TYPE: Tibet. Yadong Xian, Chun-Pei-River. on rock. 2900 m. Zang Mu 29 V '75, n° 71 (holotype, IFSBH!; isotype, NY!).

DISTRIBUTION: CHN (*W.-X. Xu 42*, ALTA); MWI (*M.J. Wiggington M1298*, MA).

37. Grimmia mariniana Sayre, Bryologist 58: 323. 1955. PROTOLOGUE: [U.S.A.] California, Marin County. On exposed, north-facing rocks, summit of Mt. Tamalpais at 3500 ft. Type Specimen: W.C. Steere, 8 April 1953. In herbaria of Stanford University, the New York Botanical

Garden and the author. Exsiccati: Two further specimens from Mt. Tamalpais, as follows: L.F. Koch 2274, 10 August 1947; J.T. Howell, 3 April 1949 (communicated by Koch). Also from Mt. S. Helena, Napa County, California: L.F. Koch 2125, 7 July 1947. TYPE: California. Marin County [...] just under summit of Mt. Tamalpais [...] April 8, 1953. W.C. Steere (lectotype, here designated, NY!).

DISTRIBUTION: USA (*W.C. Steere s.n.*, NY).

38. Grimmia molesta J. Muñoz, Ann. Missouri Bot. Gard. 86: 152. 1999. PROTOLOGUE and TYPE: Peru. Arequipa: Arequipa, am Weg nach Puno, Hegewald & Hegewald 5480 (holotype, MO!; isotypes, H!, MA!, NY!).

DISTRIBUTION: PER (*P. Hegewald & E. Hegewald 5480*, MA).

39. Grimmia mollis Bruch & Schimp., Bryol. Europ. (fasc. 42) 3: 133, tab. 253. 1849. *Grimmia mollis* f. *emersa* Lorentz, Moosstud.: 93. 1864, nom. inval. *Hydrogrimmia mollis* (Bruch & Schimp.) Loeske, Stud. Morph. Syst. Laubm.: 108. 1910. PROTOLOGUE: [Norway] In alpe Goustafjeld provinciae norvegiacae Tellemarken H. Holmgren legit et misit. TYPE: (BM not seen).

Grimmia mollis f. *xerotrophica* Györffy, in Bauer, Musci europaei exsiccati, Schedae und Bemerkungen zur 33 Serie, n° 1612. 1924. *Hydrogrimmia mollis* f. *xerotrophica* (Györffy) Podp., Consp. Musc. Eur.: 274. 1954. PROTOLOGUE and TYPE: [Slovakia] Hohe Tatra (Montes Carpathi): Klein Kolbachtal, Fünfsee, auf Granitplatten des Abflusses des zweiten Sees in den dritten, 1980 m s.m., 4.Sept.1922 legit I. Györffy (lectotype, here designated, MO!; isolectotype, FH!).

Grimmia evansii E. Britton, Rhodora 1: 148, pl. 7. 1899. PROTOLOGUE: [U.S.A.] Tuckerman's Ravine, Mt. Washington, N.H. [New Hampshire], alt. 1,230 m. July 30, 1890. Collected by Alexander W. Evans of Yale University, [...] TYPE: Tuckerman's Ravine, Mt. Washington, N.H. A.W.E. July 30, 1890 (holotype, NY!).
Synonymized by Allen (1995: 160).

DISTRIBUTION: AUT (*J. Breidler s.n.*, JE); BEL (*V. Payot s.n.*, JE); BGR (*L. Meinunger 14306*, JE); CAN (*T.A. Hedderson 2041*, FH); CHE (*T. Herzog 352*, JE); DEU (*W. Kruger s.n.*, JE); ESP (*V. Allorge s.n.*, PC); FRA (*J.P. Hébrard 4-9882*, MO); ITA (*G. Venturi s.n.*, FH); MAR (*M.J. Cano et al. s.n.*, MA); NOR (*R.W. Hartman s.n.*, H); POL (*H. Bednarek-Ochyra & R. Ochyra 127/90*, MO); SFE (*C. Wright 73*, FH); SIB (*M. Ignatov 0/466*, MA); SVK (*I. Györffy s.n.*, MO); SWE (*H. Möller s.n.*, MO); USA (*F.J. Hermann 20740*, RM).

40. Grimmia montana Bruch & Schimp., Bryol. Europ. (fasc. 25–28) 3: 128, tab. 250. 1845. *Guembelia montana* (Bruch & Schimp.) Hampe, Bot. Zeitung (Berlin) 4(8): 125. 1846. *Grimmia holmiensis* Lindb. ex Hartm., Handb. Skand. Fl. ed. 7: 374. 1858, nom. inval., pro synon. PROTOLOGUE: In rupibus schistosis, basalticis et porphyraceis: prope Angers Galliae occidentalis (Guépin) [France], in valle badensi Murgthal dicta (Alex. Braun) [Germany], in Hassia superiore prope Biedenkopf (Bruch) [Germany], in Vogesi monte Donnersberg (Gümbel)

[Germany], prope Christianiam (Blytt) [Norway], in alpe Dovrefjeld Norvegiae (W.P. Sch.) [Norway]. TYPE: Felsen auf dem Donnersberg. Gümbel. April 1843 (lectotype, designated by Cao & Vitt (1986), BM!).

Grimmia laxa Müll. Hal., Bot. Zeitung (Berlin) 5: 801. 1847. *Guembelia laxa* (Müll. Hal.) Müll. Hal., Syn. Musc. Frond. 1: 771. 1849. PROTOLOGUE: Mexico in monte Orizabae, ubi legerunt Deppe et Schiede. Cf. eor. Collect. No. 1070a. TYPE: Mexico in monte Orizabae. Deppe et Schiede lgre. (lectotype, designated by Muñoz (1998b: 389), PC!, isolectotype, PC-Besch!).
Synonymized by Muñoz (1998b: 389).

Grimmia schiedeana Müll. Hal., Bot. Zeitung (Berlin) 13: 765. 1855. *Grimmia schiedei* Müll. Hal. ex Kindb., Enum. Bryin. Exot.: 61. 1888, nom. inval., orthogr. err. PROTOLOGUE: Mexico: Deppe et Schiede in locis haud designatis. TYPE: Mexico. Leg. Deppe et Schiede (lectotype, designated by Muñoz (1999a: 152), BM!; isolectotype, PC!).
Synonymized by Muñoz (1999a: 152).

Grimmia fragilis Schimp., Syn. Musc. Eur. ed. 2: 257. 1876, nom. illeg., non F. Weber, 1804. *Grimmia montana* var. *fragilis* (Schimp.) Loeske, Laubm. Eur. Part I: 99. 1913. PROTOLOGUE: [Portugal] ad rupes graniticas in alpestribus provinciae lusitanicae Beira, ubi cl. Welwitsch anno 1848 legit. TYPE: ad rupes granit. in alpestribus Prov. Beira. Aug. 1848. L. Welw. (lectotype, designated by Muñoz (1998b: 389), BM!).
Synonymized by Muñoz (1998b: 389).

Guembelia tenella Müll. Hal., Bot. Centralbl. 44: 388. 1890. *Grimmia tenella* (Müll. Hal.) Kindb., Enum. Bryin. Exot., Suppl. 2: 107. 1893. PROTOLOGUE: [U.S.A.] Idaho, Coeur d'Alene, 6. Aug. 1888. Dr. Röll [...]. TYPE: 6 Aug 1888 Coeur d'Alene. Dr. Julius Röll. Idaho (lectotype, designated by Muñoz (1998b: 390), G!; isolectotypes, H-BR!, JE!, S!).
Synonymized by Muñoz (1998b: 389).

Grimmia montana var. *longifolia* Cardot, in Gasilien, Rev. Bryol. 21: 24. 1894. *Grimmia montana* f. *longifolia* (Cardot) Podp., Consp. Musc. Eur.: 281. 1954. PROTOLOGUE: [France] Granit découvert, sommet de Pierre-sur-Haute (1600 m. alt.).- Fertile. TYPE: Auvergne, Pierre sur Haute 1884 (lectotype, designated by Muñoz (1998b: 390), TRH!).
Synonymized by Muñoz (1998b: 389).

Grimmia arctophila Kindb., Rev. Bryol. 23: 18. 1896. PROTOLOGUE: Amer. Greenland: Berggren. Canada, Cypress hills: Macoun. TYPE: Plantae in itineribus Suecorum polaribus collectae. Groenlandia boreal.: Claushavn 1870. S. Berggren (lectotype, designated by Muñoz (1999a: 152), S!).
Synonymized by Muñoz (1999a: 152).

Grimmia pilosissima Herzog, Rev. Bryol. 38: 12. 1911. PROTOLOGUE: [Italy] Sardaigne; sur les rochers du versant oriental de la crête du Monte Gennargentu, vers 1750 m., 2 juin 1906. TYPE: An Felsen der Ostseite der Punta la Marmora (Mte. Gennargentu). Leg. Th. Herzog 2.VI.06. ca. 1750m (holotype, JE!).

Grimmia montana var. *abnoba* H. Schmidt, Mitt. Bad. Landesvereins Naturk. Naturschutz Freiburg 2: 121, fig. 6. 1927. PROTOLOGUE: [Germany] Schauinsland: Gneisfelsen nahe dem höchsten

Gipfel und bei den unteren Häusern von Hofsgrund (S.). TYPE: Süd. Schwarzwald: Felsen nahe südweste dem Schauinslandgipfel. Schmidt (lectotype, designated by Muñoz (1998b: 390), JE!). Synonymized by Muñoz (1998b: 389).

Grimmia montana f. *submutica* J.E. Zetterst. ex H. Möller, Ark. Bot. 26A(2): 31. 1934. PROTOLOGUE: [Sweden] In saxis juxta Husbyborg prope Upsaliam 15/5 1855. [Zetterstedt, Grimmiae et Andreaeae exsiccatae, nº 21c]. TYPE: In saxis juxta Husbyborg prope Upsaliam 15/5 1855. [Zetterstedt, Grimmiae et Andreaeae exsiccatae, nº 21c] (lectotype, designated by Muñoz (1998b: 390), H-BR!).
Synonymized by Muñoz (1998b: 389).

DISTRIBUTION: AND (*C. Casas s.n.*, BCB); AUT (*H. Gander s.n.*, TRH); BEL (*A. Cornet s.n.*, NY); CAN (*J. Macoun s.n.*, FH); DEU (*S. Caspari & J. Heinrichs 2526*, MO); ESP (*J.L. Rupidera 49*, SALA); FRA (*F.M.L. Corbière s.n.*, FH); GBR (*C.H. Binstead s.n.*, FH); GRL (*W.C. Steere 62-621*, NY); ISL (*B. Jóhannsson 11813*, ICEL); ITA (*F. Kern s.n.*, WRSL); MEX (*A.J. Sharp 4289*, TENN); NOR (*J.E. Zetterstedt s.n.*, FH); POL (*K.G. Limpricht s.n.*, FH); PRT (*F.M.J. Welwitsch 53*, BM); SWE (*S. Arnell s.n.*, MA); TUR (*E. Nyholm & A.C. Crundwell 674*, F); TZA (*G.H.S. Wood 1730*, MA); USA (*A.T. Whittemore 4277*, MO).

41. Grimmia muehlenbeckii Schimp., Syn. Musc. Eur.: 212. 1860. *Grimmia trichophylla* subsp. *muehlenbeckii* (Schimp.) Boulay, Musc. France 1: 380. 1884. *Grimmia trichophylla* var. *muehlenbeckii* (Schimp.) Husn., Muscol. Gall.: 135. 1887. *Dryptodon muehlenbeckii* (Schimp.) Loeske, Stud. Morph. Syst. Laubm.: 111. 1910. PROTOLOGUE: [Austria] in rupibus graniticis regione subalpinae: prope Trafoi Tyrol merid. ubi aestate 1840 [...] Mühlenbeck reperi. TYPE: Trafoi. Jul. 40 (holotype, BM!).

Grimmia rugulosa Lindgr., Bot. Not. 1845: 88. 1845, nom. nud. ORIGINAL SPECIMEN: [Sweden] Upsaliae 1834. Lindgr. (BM!).

Grimmia trichophylla var. *septentrionalis* Schimp., Coroll. Bryol. Eur.: 47. 1856. PROTOLOGUE: [Sweden] In sylvaticis circa Holmiam et Upsaliam abundanter legi. TYPE: Upsala. Schimper (syntype ["Upsala"] JE!).

Grimmia hermannii Crum, Michigan Bot. 13: 171. 1974. PROTOLOGUE: [U.S.A.] On boulder in woods below Falls, Manganese River gorge, Copper Harbor, Keeweenaw [sic] County, F.J. Hermann 25446, July 14, 1973, type in herb. MICH, duplicates in herb. US, NY, DUKE, F, CAN, WTU. TYPE: Upper Peninsula of Michigan. Keweenaw County. On boulder in woods below Falls, Manganese River gorge, Copper Harbor. F.J. Hermann 25446. July 14, 1973 (isotypes, CANM-226826!, F!, FH!).
Synonymized by Muñoz (2000c).

DISTRIBUTION: CAN (*G.R. Brassard 10990*, NY); CHE (*P.G.M. Rhodes 267*, MO); DEU (*J. Heinrich & C. Schmidt JH3740,* GOET); ESP (*C. Casas s.n.*, MA); FIN (*S.O. Lindberg s.n.*, MO); FRA (*R. Spruce s.n.*, NY); GEO (*G. Woronow 77a*, PC); HUN (*L. Marstaller s.n.,* JE); ITA (*G. Venturi s.n.*, FH); LVA (*J. Treboux s.n.*, PC); NOR (*P. Størmer s.n.*, BCB); POL (*R. Ochyra s.n.*, MO);

PRT (*C. Sérgio & S. Fontinha s.n.*, LISU); SIB (*E.A. Selivanova-Gorodkov 2460*, LE); SUN (*M. Ignatov & A. Bezgodov 244*, NY); SWE (*J.E. Zetterstedt s.n.*, H); TUR (*P. Uotila 35659*, H-BR); USA (*B. Allen 10358*, MO).

42. Grimmia navicularis Herzog, Beih. Bot. Centralbl. 26(2): 65. 1909. PROTOLOGUE: Bolivia: An Felsen in der Blockhalde über dem Tunarisee (Prov. Cochabamba), ca. 4400 m; Januar, 08, steril; an Felsen bei der Abra de San Benito (Prov. Cochabamba), ca. 3900 m; Januar, 08, c. thecis vetustis. TYPE: Flora von Bolivia. An Felsen des Abra de San Benito (Prov. Cochabamba) ca. 3900m. T. Herzog. Januar, 08 (lectotype, designated by Deguchi (1987: 27), JE!; syntypes ["Cochabamba: über dem Tunarisee"] JE!, PC!).

Grimmia paramophila Broth., in Meyer, In den Hoch-Anden von Ecuador: 518. 1907, nom. nud. ORIGINAL MATERIAL: Ecuador. Chimborazo: Paramo-region, bei 5000 m. Seehöhe. N.N.W. Seite. 1903.VII leg. Prof. Dr. Hans Meyer.

Grimmia biplicata Broth., in Meyer, In den Hoch-Anden von Ecuador: 518. 1907, nom. nud.

Grimmia chilensis Thér., in Herzog, Darwiniana 11: 217. 1957. PROTOLOGUE: Rep. Argentina: Cerro López, 1700 m, leg. A. Donat, nº 114. TYPE: Parque Nac. Arg. Nahuelhuapi no. 114. Cerro Lopez, 1700 m. Leg. A. Donat, 1.XI.36 (lectotype, designated by Muñoz (1999a: 152), JE!; isolectotype, JE!).
Synonymized by Deguchi (1987: 28).

DISTRIBUTION: ARG (*A. Donat 114*, JE); BOL (*M. Lewis 79-1731*, F); COL (*A.M. Cleef 6965*, NY); ECU (*H. Meyer 5478*, H); PER (*E. Hegewald 6001*, MO); VEN (*D. Griffin III et al. 1318*, FLAS).

43. Grimmia nepalensis Mitt., J. Linn. Soc., Bot. Suppl. 1: 46. 1859. PROTOLOGUE: In Nepal orient. reg. alp., J.D. Hooker! (No. 248, ex parte). TYPE: 248 partly. E Nepal alt. 14500 pd. J.D. Hooker (holotype, NY!).

Grimmia inflectens Mitt., J. Linn. Soc., Bot. Suppl. 1: 43. 1859. PROTOLOGUE: In Nepal orient. reg. alp., J.D. Hooker! (No. 248, ex parte). TYPE: East Nepal, 14500 ped. 248 in part. J.D. Hooker (lectotype, here designated, NY!).

DISTRIBUTION: IND (*J.D. Hooker 325*, NY); NPL (*J.D. Hooker 248*, NY).

44. Grimmia nivalis Kindb., Bull. Torrey Bot. Club 17: 271. 1890. PROTOLOGUE: [U.S.A.] On rocks on the summit of the Gold Range, north of Griffin Lake, B.C. Alt. 7,000 feet. Aug. 10, 1889. TYPE: N. Amer. Brit. Columbia, Gold Range, rocks, 7000 f. 10/8 89 J. Macoun (lectotype, designated by Muñoz (1998b: 392), S!; isolectotypes, CANM-198090!, H-BR!).

Grimmia papillinervis Kindb., Rev. Bryol. 34: 89. 1907. PROTOLOGUE: Canada, Brit. Col., Skagit summit: J.M. Macoun 1905. TYPE: Canada, Brit. Col., Skagit summit, dry rocks. 18/7 05. J.M. Macoun (holotype, Muñoz (1998b: 392), S!; isotype, FH!).
Synonymized with *G. tenerrima* by Allen (1995: 162–164), but see Muñoz (1998b: 392).

DISTRIBUTION: CAN (*J.M. Macoun s.n.*, S); USA (*J.T. Howell s.n.*, FH).

45. Grimmia nutans Bruch, in F.A. Müll., Flora 12: 390, tab. 8. 1829. PROTOLOGUE: [Turkey] Auf Felsen bei Smyrna. TYPE: Smyrna (lectotype, here designated, H-SOL!; isolectotype, NY!).

Grimmia verticillatula Thér. & Trab., Bull. Soc. Hist. Nat. Afrique N. 21: 29. 1930. PROTOLOGUE: [Morocco] in rupibus arenaceis Atlantis Majoris: in ditione Glaoua supra pagum Zerekten, ad alt. 1400–1600 m (Dr. R. Maire, 1926). TYPE: Algérie [sic, the locality is in Morocco]. Grand Atlas. Zerekten 1600 m. Legit R. Maire n° 2055 (lectotype, here designated, PC!).

Grimmia meteorae C.C. Towns., in K. Tan, The Davis & Hedge Festschrift: 48. 1989. PROTOLOGUE and TYPE: Greece. Thessaly: on damp stony bank near the Varlaam Monastery, the Meteora, near Kalabaka, apparently calcareous, 10 May 1987, C.C. Townsend 87/125 (E, holotype; Herb. Townsend, isotype) (isotype, NY!).

DISTRIBUTION: CYP (*H. Greven s.n.*, MA); GRC (*C.C. Townsend 87/125*, NY); MAR (*R. Maire s.n.*, PC); TUR (*Anonymous s.n.*, H-SOL).

46. Grimmia obtusifolia C.H. Gao & T. Cao, Acta Bot. Yunnan. 3: 394, tab. 4 fig. 10–16. 1981. PROTOLOGUE: [Tibet] Xizang: Districtus Shuanghu Xian, ad rupes, alt.s.m. 4850 m, 13.VII.1978, Lang Kai-Yong 1347 (Typus in ISPH, Isotypus in HP); Bangkog Xian, ad rupes, alt.s.m. 4800 m, 5.VI.1976, Lang Kai Yong 1232. TYPE: Tibet, Shuanghu Xian, Lang 1347 (holotype, IFSBH!; paratypes, IFSBH!, MO! ["Lang 1232"]).

DISTRIBUTION: CHN (*P.L. Redfearn Jr. 35436*, MO).

47. Grimmia ochyriana J. Muñoz, Nova Hedwigia 66: 235–240. 1998. PROTOLOGUE and TYPE: Népal. Près du Camp de Base, direction de Lobuje, 5100 m, 31-V-1952, Zimmermann 558 (holotype, G!; isotype, MA-5212!).

Grimmia mexicana Greven, Bryologist 102: 432, fig. 2. 1999. PROTOLOGUE: Mexico. Puebla. Pico de Orizaba, track to Piedra Grande, elev. 3,870 m, sunny S-facing andesite rock, 4 Dec. 1997. H.C. Greven M100 (Holotype, BM; isotypes ENCB, FH, MEXU, MICH, NY). TYPE: Mexico. Puebla: volcán Orizaba, 2.5 km below Piedra Grande, 3890m, 4 Dec. 1997. H.C. Greven M100 (isotypes, MEXU!, NY!).

NOTE: The holotype is not deposited in BM yet (L.T. Ellis, *pers. comm.*).

DISTRIBUTION: GTM (*A.J. Sharp 5428*, TENN); MEX (*A.J. Sharp 4270*, TENN); NPL (*A. Zimmermann 558a*, MA).

48. Grimmia olneyi Sull., in Sull. & Lesq., Musc. Bor.-Amer.: 32. 1856. PROTOLOGUE and TYPE: [U.S.A.] in rupibus maritimis Rhodensis Insulae (cl. S.T. Olney legit). [Sullivant & Lesquereux, Musci borali-americani exsiccati n° 141] (lectotype, here designated, FH!; isolectotypes, F!, MO!, NY!).

Grimmia austinii Kindb., Rev. Bryol. 23: 19. 1896 ["Austini"]. PROTOLOGUE: Amer. Unit. St., N. Jersey: Austin, com. Macoun. TYPE: N. Amer., New Jersey: Austin. com. Macoun (lectotype, here designated, S!).

DISTRIBUTION: CAN (*J.M. Gillet 17125*, CANM); USA (*A.J. Sharp 35255*, MO).

49. Grimmia orbicularis Bruch ex Wilson, Engl. Bot. Suppl. 4: 2888. 1844. *Guembelia orbicularis* (Wilson) Hampe, Bot. Zeitung (Berlin) 4(8): 125. 1846. PROTOLOGUE: [United Kingdom] Found on Ormeshead in 1826 by the writer of this article, and subsequently by F.K. Eagle, Esq., on St. Vicent's Rocks. For recent specimens from the latter place we are indebted to Mr. Thwaites, who gathered them in March 1843. TYPE: St. Vicent's Rocks. Feb. 1837 (Eagle) (lectotype, here designated, BM!).

NOTE: The label of the specimen selected as lectotype by Geissler & Maier (1995) reads: "Grimmia orbicularis *Bruch*/Grimmia africana/(Fissid. africanus Hedw. Sp. M.)/St. Vicent's Rocks near Bristol/(Mr. Stephens)/discovered by Mr. Eagle in 1837. W.W." (BM!). It clearly states "Stephens" as collector and not "Eagle." Stephens's name did not appear in the protologue of *Grimmia orbicularis*, and hence this specimen cannot be designated as lectotype (ICBN Art. 9.13).

Grimmia sinaica Bruch & Schimp., Bryol. Europ. (fasc. 25–28) 3: 113. 1845. *Guembelia sinaica* (Bruch & Schimp.) Hampe, Bot. Zeitung (Berlin) 4(8): 124. 1846. PROTOLOGUE: [no locality given]. TYPE: in summo monte Sinai [Egypt] (lectotype, here designated, BM!; isolectotypes, BM! 4 replicates, PC-Besch!).

Grimmia mammilaris Poech, in Keil, Lotos 1: 177. 1851. PROTOLOGUE: [Czech Republic] Gemein bei St. Prokop und Kuchelbad nächst Prag. TYPE: In rupibus calcareis prope Pragam. Mat. Vere. Dr. Poech. [Poech, Musci bohemici, n° 29. 1846] (lectotype, here designated, G!).

Grimmia orbicularis var. *africana* Besch., Cat. Mouss. Alg.: 17. 1882, nom. nud. *Grimmia orbicularis* var. *bescherellei* Corb., Compt.-Rend. Assoc. Franç. Avanc. Sci. 34: 482. 1905. *Grimmia pulvinata* var. *bescherellei* (Corb.) Wijk & Margad., Taxon 9: 190. 1960. PROTOLOGUE: [Algérie, Constantine] Chettabah; près l'usine Micout, 30 déc. 1903 [Leg. R. Chudeau]. TYPE: Constantine: usine Micout, 30 déc. 1903. Leg. R. Chudeau (lectotype, here designated, CHE!; isolectotypes, CHE! 3 replicates).

Grimmia pulvinatula Müll. Hal., Bot. Jahrb. Syst. 5: 81. 1884. PROTOLOGUE: Kerguelen [Foundery branch, Nov. 1874, Naumann]. TYPE: Kerguelen. D. Naumann 1874 (lectotype, here designated, PC!).

Grimmia stolonifera Müll. Hal., Bot. Jahrb. Syst. 5: 81. 1884. PROTOLOGUE: Kerguelen [Foundery branch, Nov. 1874, Naumann]. TYPE: Kerguelens Land. Dr. Fr. Naumann. Dcber 1874 (lectotype, here designated, PC!).

Grimmia orbicularis var. *persica* Schiffn., Oesterr. Bot. Z. 47(4): 129. 1897. PROTOLOGUE: [Iran] Persiae australis prov. Faristan: in rupestribus calcareis montium ad Schiras. 1650–1700 m.s.m. 25.X.1892. c.fr. (Bornm. Iter Pers.-turc. Nr. 4448-4448b). - Persiae australis prov. Faristan: Persepolis in decliv. apricis. ±1800 m s.m. 3.XI, 1892; c.fr. (Bornm. iter Pers.-turc. Nr. 4458). - Persia borealis: ad Teheran in planitie. ±1200 m s.m. 31.I.1892; c.fr. (Bornm Iter Pers.-turc. Nr. 4452). TYPE: J. Bornmüller: Iter Persico-turcicum 1892–93. N° 4448. Persiae australis prov. Faristan: in rupestribus calcar. montium ad Schiras. 1650–1700 m.s.m. 1892.X.25 legit et determ.: J. Bornmüller (lectotype, designated by Muñoz (2000c), JE!; isolectotypes, F!, PC!). Synonymized by Muñoz (2000c).

Grimmia microglobosa Müll. Hal., Hedwigia 37: 160. 1898. PROTOLOGUE: [New Zeeland] Nova Seelandia, insula australis, North Canterbury, Ben More, ad rupes alt. 5000 ped., Nov. 1890: T.W. Naylor Beckett, qui 1892 misit. TYPE: On rocks Ben More alt. 5000' N. Canterbury. T.W.N.B. November 1890 (lectotype, here designated, H-BR!).

Grimmia orbicularis var. *brevipila* J.J. Amann, Fl. Mouss. Suisse 2: 134. 1918. *Grimmia orbicularis* f. *brevipila* (J.J. Amann) Podp., Consp. Musc. Eur.: 285. 1954. PROTOLOGUE: [Switzerland] Vaud: Saint-Saphorin, murs de vignes, 400–450 m (BH 91.6.20 et 22) (Amann). TYPE: [Switzerland] Vaud: Saint-Saphorin, 400–450 m, 18.XII.1912. Leg. Amann (BH 91.6.22) (lectotype, here designated, Z!; isolectotype, PC!; syntype ["*Amann BH 91.6.20*"], Z!).

Grimmia orbicularis var. *humilis* J.J. Amann, Fl. Mouss. Suisse 2: 134. 1918. *Grimmia orbicularis* f. *humilis* (J.J. Amann) Podp., Consp. Musc. Eur.: 285. 1954. PROTOLOGUE: [Switzerland] Vaud: rochers de poudingue au Dézaley sous Chexbres, 450–500 m (BH 91.6.26 et 28) (Amann). TYPE: [Switzerland] Vaud: rochers au Dejaley 450–500 m, 3.3.1912. Leg. Amann (BH 91.6.28) (lectotype, here designated, Z!; syntype ["*Amann BH 91.6.26*"], Z!).

Grimmia moxleyi R.S. Williams, in Holz., Musci Acrocarpi Boreali-Americani et Europaei, fasc. 24 n° 600. 1926. PROTOLOGUE and TYPE: [U.S.A.] On sandstone conglomerate, between Big Rock Creek and Devil's Punch Bowl, Los Angeles Co., Calif. Coll. George L. Moxley, No. 1141. May 30–31, 1926. Holzinger, Musci Acrocarpi Boreali-Americani et Europaei, n° 600 (lectotype, designated by Muñoz (2000c), NY!; isolectotypes, FH!, H-BR!, MO! 3 replicates). Synonymized by Muñoz (2000c).

DISTRIBUTION: ARM (*V.A. Manakyan 3783*, LE); ATF (*F.C. Naumann s.n.*, PC); AUS (*R.G. Coveny 15709*, MO); AZE (*Alexeenko s.n.*, LE); BEL (*J.L. De Sloover 24805*, MA); CHE (*J. Amann s.n.*, Z); CZE (*J. Podpěra s.n.*, MA); DEU (*L. Meinunger 18391*, JE); DZA (*L. Trabut 2123*, PC); EGY (*J.-P. Frahm s.n.*, NY); ESP (*M.J. Elías 336*, SALA); FRA (*H.N. Dixon 10*, MO); GEO (*A.H. Brotherus & V.F. Brotherus 131*, H-SOL); GRC (*V. Schiffner 77*, FH); HUN (*R. Marstaller s.n.*, JE); IND (*W. Koelz 7304A*, MO); IRN (*J. Bornmüller 4448*, JE); IRQ (*H.F.v. Handel-Mazzetti 1214*, FH); ITA (*K. Loitlesberger s.n.*, MA); JOR (*W. Frey & H. Kürschner 4606*, MA); MAR (*P. Davis & J. Davis D48490A*, NY); MKD (*T. Herzog 625*, JE); NZL (*T.W.N. Beckett s.n.*, H-BR); OMN (*H. Kürschner 99-198*, MA); PAK (*G. Kerstan 58/2*, JE); POL (*H. Greven s.n.*, MA); PRT (*C. Sérgio et al. 31.3b*, BCB); SAU (*W. Frey & H. Kürschner 3136*, MA); SUN (*V. Vašák s.n.*, NY); SYR (*H.K. Haussknecht 40*, JE); TKM (*U. Mamatkulov 102*, MO); TUN (*J.-P. Frahm 844510*, MO); TUR (*E. Yucei 80/1*, MO); USA (*F.A. MacFadden 3809*, MO); UZB (*C.C. Townsend 86/193*, ALTA); YUG (*M. Sabovljević s.n.*, MA).

50. Grimmia ovalis (Hedw.) Lindb., Acta Soc. Sci. Fenn. 10: 75. 1871. *Dicranum ovale* Hedw., Descr. Micr.-Anal. Musc. Frond. 3: 81, tab. 34 fig. 1–6. 1792, nom. inval. *Bryum ovale* Hoffm., Deutschl. Fl. 2: 35. 1795, nom. inval. *Dicranum ovatum* Sw., Kongl. Vetensk. Acad. Nya Handl. 16: 243. 1795, nom. inval. *Dicranum ovale* Hedw., Sp. Musc. Frond.: 140. 1801. *Trichostomum ovatum* P. Beauv., Prodr. Aethéogam.: 91. 1805, nom. illeg., incl. typ. spec. prior. *Dicranum ovatum* Brid., Muscol. Recent. Suppl. 1: 214. 1806, nom. illeg., incl. typ. spec. prior. *Grimmia elliptica* Funck, Deutschl. Moose: 16, tab. 11 fig. 1. 1820, nom. illeg., incl. typ. spec. prior. *Trichostomum ovale* (Hedw.) Roehl. ex Steud., Nomencl. Bot. 2: 99 [also 421]. 1824, nom. inval., pro synon. *Campylopus ovalis* (Hedw.) Wahlenb., Fl. Suec. 2: 748. 1826. *Dryptodon ovatus* (Hedw.) Brid., Bryol. Univ. 1(1): 202. 1826. *Grimmia commutata* Huebener, Muscol. Germ.: 185. 1833, nom. illeg., incl. typ. spec. prior. *Dryptodon ellipticus* (Funck) Hartm., Handb. Skand. Fl. ed. 3: 271. 1838, nom. illeg., non Brid. 1826. *Guembelia elliptica* (Funck) Hampe, Bot. Zeitung (Berlin) 4: 125. 1846. *Guembelia ovalis* (Hedw.) Müll. Hal., Syn. Musc. Frond. 1: 774. 1849. *Guembelia commutata* (Huebener) Rabenh., Krypt. Fl. Sachsen 1: 444. 1863, nom. illeg., incl. typ. spec. prior. *Dryptodon ovalis* (Hedw.) Hartm. ex H. Möller, Ark. Bot. 26A(2): 14. 1934, nom. inval., pro synon. PROTOLOGUE: In rupestribus montosis Saxoniae [Germany], (Thuringiae prope Isenacum [Germany], in granite montis piniferi Franconiae, Austriae prope Engelhardszell). TYPE: [no locality given] (lectotype, designated by Geissler & Maier (1995: 502), G!).

Bryum brevicaule Vill., Hist. Pl. Dauphiné 3(2): 871. 1789, nom. inval. PROTOLOGUE: [France] dans les Alpes, à Orciere, Saint-Leger, dans le Champfaur, en Valgaudemar & ailleurs.

Grimmia ovata F. Weber & D. Mohr, Naturh. Reise Schweden: 132, tab. 2 fig. 4a–c. 1804. *Campylopus ovalis* var. *ovatus* (F. Weber & D. Mohr) Wahlenb., Fl. Suec. ed. 2, 2: 776. 1833. *Dryptodon ovatus* (F. Weber & D. Mohr) Hartm., Handb. Skand. Fl. ed. 3: 271. 1838, nom. illeg., non Brid. 1826. *Grimmia ovata* var. *euovata* Loeske, Laubm. Eur. Part I: 113. 1913, nom. inval. PROTOLOGUE: [Norway] Habitat in rupibus prope Holmiam. Accepimus quoque e Norvagia [sic]. TYPE: Holmiae (lectotype, designated by Muñoz (1999a: 159), BM!).

Grimmia canescens Schleich., Neues J. Bot. 1(2): 198. 1805, nom. nud. *Grimmia nigricans* Lam. & DC., Fl. Franç. ed. 3 [original issue] 2: 459. 1805. *Dryptodon ovatus* var. *nigricans* (Lam. & DC.) Brid., Bryol. Univ. 1(1): 179. 1826, nom. inval., pro synon. PROTOLOGUE: Cette espèce m'a été envoyée des Alpes voisines du Léman [Switzerland]: elle a été trouvée dans les Pyrénées [France/Spain], par le C. Dufour, et dans les montagnes d'Auvergne [France], par le C. Dubois. TYPE: Pyrenées (lectotype, here designated, G!).

Grimmia ventricosa Schleich., Cat. Pl. Helv. ed. 4: 40. 1821, nom. nud. *Grimmia commutata* var. *ventricosa* Lesq., Mém. Soc. Sci. Nat. Neuchâtel 3(3): 27. 1846, nom. nud. ORIGINAL MATERIAL: [Helvetia]. ORIGINAL SPECIMEN: Helvetia. Schleicher [FH specimen labeled as "Europe"] (FH!, NY!, PC!).

Grimmia commutata f. *epilifera* J.E. Zetterst., Revis. Grimm. Scand.: 90. 1861. *Grimmia commutata* var. *imberbis* Lindb. ex H. Möller, Ark. Bot. 26A(2): 17. 1934, nom. inval., pro synon. *Grimmia commutata* var. *mutica* J.E. Zetterst. ex H. Möller, Ark. Bot. 26A(2): 17. 1934, nom. inval., pro synon. PROTOLOGUE: [Sweden] In saxis apricis Sueciae mediae hic illic occurrit, v.c. in Uplandia, Vestrogothia boreali, Dalia et Vermlandia [...]. TYPE: In saxis juxta Gottsunda prope Upsaliam. 5-V-1858. J.E. Zetterstedt (lectotype, designated by Muñoz (1999a: 159), H!). Synonymized by Muñoz (1999a: 159).

Grimmia cossonii Besch., Cat. Mouss. Alg.: 18. 1882. PROTOLOGUE: [Algeria] Djebel Ksel, près de Géryville, 1,307 mèt., mai 1856 (Coss.[Cosson, E.S.-C.]). TYPE: Flore de Algérie. Djebel Ksel pr. Geryville. n° 1762 ex herb. Bescherelle. Legit Cosson 30 mai 1856 (lectotype, designated by Muñoz (1999a: 159), PC!).
Synonymized by Muñoz (1999a: 159).

Grimmia ovata var. *gracilis* Röll, Hedwigia 32: 265. 1893, nom. nud. ORIGINAL MATERIAL: [U.S.A.] Idaho: Coeur d'Alène (1163). ORIGINAL SPECIMEN: Coeur d'Alene. Idaho. 6/888 J. Röll (H-BR!).

Grimmia bernoullii Müll. Hal., Bull. Herb. Boissier 5(3): 200. 1897. PROTOLOGUE: [Guatemala] Quezaltenango; Coll. N° 115 partim [leg. Bernoulli & Cario]. TYPE: Quezaltenango, leg. Bernoulli & Cario (lectotype, designated by Muñoz (1999a: 159), PC!; isolectotype, GOET!).
Synonymized by Muñoz (1999a: 159).

DISTRIBUTION: AFG (*K. Holmen s.n.*, ALTA); AND (*C. Sérgio s.n.*, LISU); ARM (*V. Manakyan 4610a*, ERE); AUT (*C.C. Townsend 92/552*, MA); BEL (*C. Delogne s.n.*, FH); BGR (*V. Valchev & V. Nikolov s.n.*, MO); CAN (*F.A. MacFadden 3849*, RM); CHE (*J.C. Schleicher s.n.*, NY); CHN (*B.C. Tan 93-1242*, FH); CYP (*F. Unger s.n.*, JE); DEU (*H. Müller s.n.*, FH); DZA (*J.-P. Frahm 844540*, MA); ESP (*A.C. Crundwell 377*, NY); FIN (*V.F. Brotherus s.n.*, JE); FRA (*J.-P. Frahm 935072*, MA); GEO (*V.F. Brotherus 229*, H-SOL); GRC (*H.K. Haussknecht s.n.*, JE); GTM (*C.G. Bernoulli & R. Cario s.n.*, PC); HUN (*L. Meinunger 7404*, JE); ISL (*B. Jóhannsson 33267*, ICEL); IND (*J.F. Duthie 10830*, JE); IRN (*P. Goldblatt 822*, MO); ITA (*M. Fleischer s.n.*, FH); KAZ (*B. Allen 10750*, MO); KGZ (*V.F. Brotherus s.n.*, H-BR); LKA (*Anonymous s.n.*, BM); MAR (*L. Trabut 1713*, PC); MEX (*A.J. Sharp 784*, TENN); MKD (*T. Herzog 570*, JE); NOR (*S.C. Sommerfelt s.n.*, H); PAK (*C. Jaggi MYQ123*, MO); POL (*K.G. Limpricht s.n.*, FH); PRT (*C. Aedo s.n.*, MA); SIB (*M. Ignatov 0/463*, MA); SUN (*V.G. Onipchenko 48/95*, NY); SVK (*Z. Pilous s.n.*, FH); SWE (*C.G. Myrin s.n.*, BM); TUR (*H.K. Haussknecht 60*, JE); TZK (*P. Polyakov s.n.*, NY); UKR (*D. Zerov s.n.*, NY); USA (*B. Allen 17892*, MO); UZB (*L. Nazarenko s.n.*, LE).

51. Grimmia percarinata (Dixon & Sakurai) Noguchi ex Deguchi, J. Sci. Hiroshima Univ., Ser. B, Div. 2, Bot. 16: 198. 1979 ["1978"]. *Didymodon percarinatus* Dixon & Sakurai, in Sakurai, Bot. Mag. (Tokyo) 50: 516. 1936. PROTOLOGUE: [Japan] Kiushiu: Prov. Ohsumi, Insula Yakushima, mt. Kuromi, ad rupes (Leg. Y. Doi. Typus in Herb. Sakurai Nr. 4329. 5 Aug. 1934). TYPE: Ins. Yakushima, Kiusiu, Japan. Coll. Y. Doi. Date Aug. 1934. Type (holotype, Deguchi (1978: 198), MAK not seen; isotype, BM!).

Grimmia lurida Wilson, Hooker's J. Bot. Kew Gard. Misc. 9: 323. 1857, nom. nud. ORIGINAL MATERIAL and SPECIMENS: Ceylon [Sri Lanka], Gardner n° 329 [...] Sikkim-Himalaya temperata; Lachen [India], alt. 10,000 ped., J.D.H. [J.D. Hooker n° 285] (BM!, FH!).
DISTRIBUTION: IND (*J.D. Hooker 285*, BM); JPN (*F. Mizokuti 59*, NICH).

52. Grimmia pilifera P. Beauv., Prodr. Aethéogam.: 58. 1805. PROTOLOGUE: [U.S.A.] Etats-Unis d'Amerique. TYPE: (G? not found).

Grimmia pensilvanica Schwägr., Sp. Musc. Frond. Suppl. 1(1): 91, tab. 25. 1811. *Grimmia pennsylvanica* Schwägr. ex Sull. & Lesq., Musc. Bor.-Amer.: 32. 1856, nom. inval., orthogr. err. PROTOLOGUE: [U.S.A.] Lecta in Pensilvaniae rupibus a rev. Mühlenbergio. TYPE: (lectotype, designated by Cao & Vitt (1986: 144), G not seen).
Synonymized by Cao & Vitt (1986: 144).

Grimmia arizonae Renauld & Cardot, Rev. Bryol. 19: 85. 1892. PROTOLOGUE: [U.S.A.] Ariz. [Arizona]. TYPE: Am. sept. Fort Grand, Arizonae. Leg. La Forêt (lectotype, designated by Muñoz (1999a: 161), PC!; isolectotypes, FH!, NY!).
Synonymized by Muñoz (1999a: 161).

Grimmia tenax Müll. Hal., Nuovo Giorn. Bot. Ital., n.s. 3: 109. 1896. PROTOLOGUE: [China] In monte medio Si-ku-tziu-san. TYPE: China interior, provincia Schen-si mer. in medio monte Si-Ku-tzui-san [sic] (lectotype, designated by Deguchi (1978: 201), FI not seen; isolectotype, PC!).
Synonymized by Deguchi (1978: 201).

Grimmia atroviridis Cardot, Bull. Herb. Boissier, sér. 2, 8(5): 333. 1908. PROTOLOGUE: [South Korea] Corée: Pomasa (n. 209); Hoang-hai-to (n. 658); île Quelpaert (n. 91). TYPE: Corée: Pomasa, *Faurie 209* (lectotype, designated by Deguchi (1986: 327), PC!; syntypes ["*Faurie 91*"] H-BR!, NY!).

Grimmia arsenei Cardot, Rev. Bryol. 40: 37. 1913. PROTOLOGUE: [Mexico] État de Michoacan: environs de Morelia (Frère Arsène, 1911). TYPE: Mexico. Vicinity of Morelia, State of Michoacán 2000 m. N° 7894 Bro. G. Arsène, Collector 3.XII.1911 (lectotype, designated by Muñoz (1999a: 161), PC!; isolectotypes, F!, PC! 2 replicates).
Synonymized by Muñoz (1999a: 161).

Grimmia hisauchii S. Okamura, J. Coll. Sci. Imp. Univ. Tokyo 38(4): 11, fig. 4. 1916. PROTOLOGUE: [Japan] Hondo: Prov. Sagami, Ikego, Dushi (Leg. Kiyotaka Hisauchi! 12.XII.1914.). TYPE: Sagami, Kamakura-cho. Dec. 12st 1914. Kiyotaka Hisauchi 12822 (isotype, TNS!).
Synonymized by Deguchi (1978: 201).

Grimmia santaritae E.B. Bartram, Bryologist 27: 60, pl. 9. 1924. PROTOLOGUE: [U.S.A.] Dry rocks faces, White House Canyon, Santa Rita Mts., Santa Cruz County, Arizona, E.B. Bartram, No. 694, February 20, 1923. TYPE: Mosses of Southern Arizona collected and distributed by Edwin B. Bartram. N° 694. Dry rock faces. White House Canyon, Santa Rita Mountains, Santa Cruz County. Altitude 4500 ft. Feb. 20-1923 (lectotype, designated by Muñoz (1999a: 161), FH!; isolectotypes, FH!, NY!).
Synonymized by Muñoz (1999a: 161).

Grimmia santaritae f. *propagulifera* E.B. Bartram, Bryologist 27: 60. 1924. *Grimmia arizonae* f. *propagulifera* (E.B. Bartram) G.N. Jones, in Grout, Moss Fl. N. Amer. 2(1): 33. 1933. PROTOLOGUE: [U.S.A., locality not explicitly stated]. TYPE: Mosses of Southern Arizona collected and distributed by Edwin B. Bartram. N° 618. Shaded bank in Patagonia Mountains, Santa Cruz County. Altitude 5000 ft. Feb. 13-1923 (lectotype, designated by Muñoz (1999a: 161), FH!).
Synonymized by Muñoz (1999a: 161).

Grimmia elatior var. *squarrifolia* Dixon & Thér., Hong Kong Naturalist, Suppl. 2: 10. 1933. PROTOLOGUE: [China] Exposed granite rock, 400–500 ft. alt., Amoy I., 8th July, 1931; coll. G.A.C. Herklots (B.8 A, type). Vertical face of rock, ibidem (B.15 B, B.16 B). TYPE: Granite rock, 400–500 ft. alt., Amoy Island. China. 8 July 1931. Coll. G.A.C. Herklots. No. 8.A (holotype, according to Cao & Vitt (1986: 144), BM not seen; isotype, PC!; paratypes, H!, MO! ["*Herklots B.15B*"]).
Synonymized by Deguchi (1978: 201).

Grimmia eurybasis Dixon & Sakurai, Bot. Mag. (Tokyo) 53: 287. 1939. PROTOLOGUE: [Japan] Kyusyu: Prov. Higo, Isyoti (Leg. K. Mayebara Typus in Herb. K. Sakurai Nr. 9506 23 Feb. 1936); ebenso Tunagi-Dorf (Nr. 11189 4 Okt. 1936). TYPE: (holotype, according to Deguchi (1978: 201), MAK not seen).
Synonymized by Deguchi (1978: 201).

Racomitrium doii Sakurai, Bot. Mag. (Tokyo) 54: 6 fig. 3. 1940. PROTOLOGUE: [Japan] Kiusiu: Prov. Ohsumi, Utinoura, Tarumi, auf Felsen (Leg. Y. Doi Typus in Herb. K. Sakurai Nr. 11716 30 Maerz 1938). TYPE: (holotype, according to Deguchi (1978: 201), MAK not seen).
Synonymized by Deguchi (1978: 201).

Grimmia kirienensis C.H. Gao, Flora Muscorum Chinae Boreali-Orientalis: 379, fig. 82. 1977. PROTOLOGUE: [China] Prov. Kirin: Districtus Linkiang Mt. Mao-ehr-shan 800 m, ad rupes, 7.IX.1963, Ch. Gao 7714 (Typus) (Typus in Herb. Inst. Sylv. et Pedol. Acad. Sinicae). Prov. Heilungkiang: Xiao-chingan-ling Yichun, ad rupes, 10.X.1950, Y.L. Chang 1440. TYPE: China. Jilin. Lijiang Xian. 7.9.1963. Gao Chien 7714 (holotype, IFSBH!).
Synonymized by Deguchi (1986: 327–328).

DISTRIBUTION: CAN (*J. Macoun s.n.*, MA); CHN (*C. Gao 1263*, ALTA); JPN (*A. Sakai 1066*, TNS); KOR (*U.J. Faurie 91*, H-BR); MEX (*A.J. Sharp et al. 3233*, TENN); MNG (*C. Tsegmed 1895*, LE); SFE (*R.E. Magill 12367*, MO); SIB (*V. Smirnov 5461*, MO); SUN (*P. Zhudov & Pokrovskaya s.n.*, LE); USA (*B. Allen 17830*, MO); VIR (*Krebs s.n.*, H-SOL).

53. Grimmia pitardii Corb., Bull. Soc. Bot. France 56: LVI. 1909. *Campylostelium pitardii* (Corb.) Maier, Candollea 53: 307. 1998. PROTOLOGUE: [Tunisia] [...] environs d'Aïn-Draham (Khroumirie). [J. Pitard & L. Corbière]. TYPE: (isotype, G not seen).

NOTE: Maier (1998) has transferred *G. pitardii* to *Campylostelium* after a detailed anatomical study. I agree with her on this taxon being odd in *Grimmia*, and that it shares some anatomical features with *Campylostelium*. Nevertheless, there are other characters shared with species of *Grimmia* subg. *Grimmia* and not with *Campylostelium*. At this moment I prefer to take a conservative approach and maintain this species in *Grimmia*.

Usmania campylopoda Laz., Dopov. Akad. Nauk Ukrajins'k. RSR 11: 1040. 1970. *Grimmia campylopoda* (Laz.) K. Saito, J. Jap. Bot. 48(6): 163. 1973. PROTOLOGUE: [Tajikistan] Habitat in terra nec non saxis terra obtectis ad meridiem Rei Publicae Sovieticae Socialisticae Tadzhikorum: in decliviis jugi Tuyun Tau, 650 m supra mare, N° 8266, 22.IV 1966. U. Mamatkulov legit (holotypus); ibidem, N° 8268 22.IV.1966, U. Mamatkulov leg (Isotypus);

reservatum publicum "Tigrovaya balka", in arenis Kashka Kum, 350 m supra mare; N° 8298, 23.IV.1966. U. Mamatkulov legit. TYPE: Tajikistan: South Saartus, Tugon-Tau, Karon *Mamatkulov 8266* (holotype, LWS not seen).
Synonymized by Abramova & Abramov (1988).

Grimmia gibbosa S. Agnew, J. Bryol. 7: 339. 1973. PROTOLOGUE: [Iraq] Kirkuk Liwa. Tuz Khurmatli bridge over Lesser Zab river, on stony calcareous soil on dry river bank, c.fr., 21 March 1958, N. Polunin s.n. (Type), (holotype in Herb. S. Agnew, isotypes in BUH, BM); Ain Dibbis, N. of Kirkuk, crevices of limestone outcrop on N.-facing slopes of low hills, c.fr., 26 December 1962, S. Agnew in BUH 598. TYPE: (not seen).
Synonymized by Abramova & Abramov (1988).

DISTRIBUTION: ESP (*R.M. Cros s.n.*, MA); FRA (*J.P. Hébrard A2378*, MO); GRC (*J. Muñoz s.n.*, MA); TUN (*L. Corbière s.n.*, PC); TUR (*A. Schäfer-Verwimp 20670/T*, MA); UZB (*U.K. Mamatkulov 8263*, LE).

54. Grimmia plagiopodia Hedw., Sp. Musc. Frond.: 78, tab. 15 fig. 6–13. 1801. *Grimmia plagiopus* Spreng., Bot. Zeitung (Regensburg) 1(5): 74. 1802, nom. inval., orthogr. err. *Schistidium plagiopodium* (Hedw.) Loeske, Laubm. Eur. Part I: 52. 1913. PROTOLOGUE: [Germany] In Saxoniae rupibus detexit Cl. Flügge. TYPE: Jena. Bei der Rasenmülh. Febr. 1798 leg. Flugge (holotype, G not seen; isotypes, GOET!, JE!).

Grimmia obtusa Brid., J. Bot. (Schrader) 1800(1(2)): 276. 1801. PROTOLOGUE: [Germany] prope Tenam [sic] lectam [Circa Jenam in Rosestein prope Kahla in solo nudo caespitose viuentem et terrâ quasi obrutam Vater, Bieber aliique legerunt (Bridel (1806: 94)]. TYPE: (B? not found).

Grimmia apocaula Hoffm. ex DC., in Lam. & DC., Fl. Franç. ed. 3 [original issue] 2: 458. 1805. PROTOLOGUE: [France]. TYPE: (G? not found).

Grimmia brandegei Austin, Bull. Torrey Bot. Club 6(7): 45. 1875. PROTOLOGUE: [U.S.A.] Colorado, Dr. Brandegee, (Rau). TYPE: Colorado. J.S. Brandegee 1874. Ex herb. E.A. Rau (lectotype, designated by Muñoz (1999a: 163), NY!).
Synonymized by Ochyra & Bednarek-Ochyra (1994: 669).

Grimmia plagiopodia var. *pilifera* Lesq. & James, Man.: 138. 1884. *Grimmia plagiopodia* var. *longipila* Coville ex Holz., Bryologist 28(3): 32. 1925, nom. inval., orthogr. err. PROTOLOGUE: [U.S.A.] Rocky Mountais, Colorado (E. Hall, Brandegee). TYPE: Rocky ms. legit Hall (lectotype, here designated, FH!; isolectotype, FH!).
Synonymized by Blom (1996: 262).

Grimmia argentea R. Br. bis, Trans. & Proc. New Zealand Inst. 27: 412, pl. 30 fig. 11. 1895. PROTOLOGUE: [New Zealand] Limestone rocks, Castle Hill, Canterbury; November, 1889. Collected by R.B. TYPE: Hab. limestone rocks near Castle Hill west Coast Road (from Christchurch). Coll. leg. Robert Brown 11/89 (lectotype, here designated, H-BR!; isolectotypes, H!, H-BR!).
Synonymized by Ochyra (1993).

Grimmia nivea Dusén, Bot. Not. 1905: 303. 1905. PROTOLOGUE: Patagonia australis in saxis. TYPE: [Argentina] Patagonia australis, Terr. Sta. Cruz in saxis in valle rivuli Aryo. Pelque. 16 Jan. a. 1905. P. Dusén (holotype, Deguchi (1984: 22), S not seen; isotypes, H-BR!, PC!). Synonymized by Deguchi (1984: 22).

Grimmia plagiopodia var. *antarctica* E.B. Bartram, Bryologist 60(2): 141. 1957. PROTOLOGUE: [Antarctica. Ruppert Coast] Mt. Marujupu: in crevices on granite near rockery, B37.1. TYPE: Mt. Marujupu: in crevices on granite near rockery. J.E. Perkins B.37-1, Nov. 27, 1940 (holotype, FH!). Synonymized by Ochyra (1993).

DISTRIBUTION: ARG (*P. Dusén 5713*, H-BR); ARM (*V. Manakyan 4822a*, ERE); ATA (*J.E. Perkins B37*, FH); CAN (*J. Macoun s.n.*, NY); CHL (*P.A. Siple 391-6*[a], FH); FRA (*R.B. Pierrot s.n.*, PC); DEU (*L. Meinunger 11382*, JE); GEO (*A.H. Brotherus & V.F. Brotherus 21*, H-SOL); HUN (*A. Boros s.n.*, PC); ISL (*B. Jóhannsson 25570*, ICEL); NZL (*A.J. Fife 6262*, NY); SUN (*A.H. Brotherus & V.F. Brotherus 7*, H-SOL); SVK (*Z. Pilous 314*, PC); UKR (*A.S. Lazarenko s.n.*, MA); USA (*C.E. Darigo 2600*, MO).

55. Grimmia poecilostoma Cardot & Sebille, nom. conserv. prop., Rev. Bryol. 28: 118, tab. 5. 1901. *Grimmia crinita × G. leucophaea* Philib., Rev. Bryol. 17: 18–19. 1890, nom. inval., published as a formula. *Grimmia cardotii* Hérib., Mém. Acad. Sci. Clermont-Ferrand, sér. 2, 14: 358. 1899, nom. inval., not accepted by author. *Grimmia tergestina* var. *poecilostoma* (Cardot & Sebille) Loeske, Laubm. Eur. Part I: 84. 1913. *Grimmia tergestina* var. *cardotii* (Hérib.) Podp., Consp. Musc. Eur.: 277. 1954, nom. inval., incl. typ. var. prior. PROTOLOGUE: [France.] Pont-de-Longue, sur les bordes de l'Ailler (Puy-de-Dôme) [Gasilien], et le Clapier de Saint-Christophe-en-Oisans (Isère) [Sèbille]. TYPE: Isère: Clapier de Saint-Christophe-en-Oisans, Sèbille (lectotype, designated by Muñoz (1999a: 164), PC!; syntype ["pont de Longe près Clermont. Auvergne. 3 Avril 1888. fre. *Gasilien*"], PC!).

NOTE: The name *G. poecilostoma* has been proposed for conservaton (Muñoz, 2000a).

Grimmia crassifolia Lindb. ex Broth., Acta Soc. Sci. Fenn. 19(12): 84. 1892. PROTOLOGUE: [Russia. Ossetia A.R.] Cauc. major occ.: Ossetia, ad rupes calcareas, siccisimas ad fl. Ardon inter Alagir et Misurtzy. TYPE: Ossetia, in valle fl. Ardon inter Alagir et Misurtsy, ad rupes aridissimas, 7/1877. Leg. V.F. Brotherus (lectotype, designated by Muñoz (1999a: 164), H-BR!; isolectotypes, H!, JE!, PC!, UPS!).
Synonymized by Muñoz (1999a: 164).

Grimmia gymnostoma Culm., Rev. Bryol. 23: 108. 1896. *Grimmia tergestina* var. *gymnostoma* (Culm.) G. Roth, Eur. Laubm. 1: 410. 1903. *Grimmia tergestina* f. *gymnostoma* (Culm.) Loeske, Laubm. Eur. Part I: 82. 1913, comb. inval., pro synon. PROTOLOGUE: [Switzerland] Découvert dans les fissures des rochers calcaires exposés à l'est sur l'Ebenalp, canton d'Appenzell, le 24 août 1896, retrouvé un peu plus tard au Hohen Kasten. TYPE: Au Kalkfelsen, Ostabhang der Ebenalp 1640 m., Kt. Appenzell. 24/8 1896. Culmann (lectotype, designated by Muñoz (1999a: 164), PC!).
Synonymized by Muñoz (1999a: 164).

Grimmia anodon var. *mutica* Broth. ex Paris, Index Bryol. ed. 2, 2: 263. 1904, nom. nud. ORIGINAL MATERIAL: [Kazakstan] As.- Alatau, Turkestan. ORIGINAL SPECIMEN: Alatau transiliensis: in valle fl. Almatinka minoris, ad rupes siccissimas reg. silv. infer., Jun 29, 1896. Leg. V.F. Brotherus. [Musci turkestanici nº 45] (H-BR!).

DISTRIBUTION: ARM (*V.A. Manakyan 5298*, LE); AZE (*Alexeenko s.n.*, LE); CHE (*P. Culmann s.n.*, PC); CHN (*K.-Y. Lang 1262*, ALTA); ESP (*R.M. Cros s.n.*, BCB); FRA (*G. Gasilien s.n.*, PC); GEO (*N.V. Chikovani s.n.*, LE); KAZ (*B. Allen 10508*, MO); KGZ (*V.F. Brotherus s.n.*, FH); MNG (*W. Schubert M18*, MO); SIB (*M. Ignatov 0/462*, MO); SUN (*V.F. Brotherus s.n.*, H-BR); TKM (*Navessky 3108D*, LE); TUR (*K. Walther 2832*, NY); UKR (*L. Partyka s.n.*, MA); USA (*L.R. Stark NV-25B*, MO).

56. Grimmia pseudoanodon Deguchi, in Inoue, Studies on Cryptogams in Southern Peru: 29, pl. 11. 1987. PROTOLOGUE: Peru. Chucuito, near Acora, Prov. Puno, Dept. Puno; alt. ca. 3900m, in rocky meadows, September 29, 1984, leg. H. Deguchi no. 30008 (holotype in TNS, isotypes in KOCH and NY). TYPE: Peru. Dept. Puno. Prov. Puno: Chucuito, near Acora; alt. ca. 3900 m, in rocky meadow. September 29, 1984. Coll. H. Deguchi 30008 (isotype, NY!).

DISTRIBUTION: ARG (*P.G. Lorentz s.n.*, JE); BOL (*M. Lewis 79-2004*, F); PER (*H. Deguchi 30008*, NY).

57. Grimmia pulla Cardot, Rev. Bryol. 36: 106. 1909. PROTOLOGUE: [Mexico] Etat de Hidalgo: Honey-station, sur troncs d'arbres, associé à *Zygodon circinatus* Sch., 1907 (n. 10423). TYPE: State of Hidalgo. On trees, Honey Station, 12, Dec., 1907 [Pringle, Plantae mexicanae, nº 10423] (lectotype, designated by Muñoz (1999a: 167), PC!; isolectotypes, CANM-197543!, FH!, JE!, NY!, PC! 2 replicates, TENN!, UC-133015!).

Grimmia pulla var. *squarrosa* Thér., Rev. Bryol. Lichénol. 5: 99. 1933. PROTOLOGUE: [Mexico] San Rafael (St.-P. nº 1857); Contreras, rochers (St.-P nº 1472). TYPE: 1857. Méxique: San Rafael, 2600 m (Etat de Mexico). Leg. Fr. Amable. 29.XII.1931 (lectotype, designated by Muñoz (1999a: 167–168), PC!; syntypes ["Contreras, *St. Pierre 1472*"], H-BR!, PC!).
Synonymized by Crum (1994: 405).

DISTRIBUTION: MEX (*A.J. Sharp 178*, TENN).

58. Grimmia pulvinata (Hedw.) Sm., Engl. Bot. 24: 1728. 1807. "Bryum trichodes hirsutie canescens, capitulis subrotundis reflexis in perbrevibus pediculis" Dill., Cat. Pl. Circa Gissam: 226. 1718, nom. inval. "Bryum orbiculare pulvinatum, hirsutie canescens, capsulis immersis" Dill., Hist. Musc.: 395, tab. 50 fig. 65 A–C. 1741, nom. inval. *Bryum pulvinatum* L., Sp. Pl. 2: 1121. 1753, nom. inval. *Leersia pulvinata* Hedw., Fund. Hist. Nat. Musc. Frond. 1: tab. 10 fig. 65; 2: 88. 1782, nom. inval. *Fissidens pulvinatus* Timm, Fl. Megapol. Prodr.: 216. 1788, nom. inval. *Afzelia pulvinata* Ehrh., Plantae Criptogamae Linneae, Decas 17, nº 163. 1791, nom. inval. *Encalypta pulvinata* Sibth., Fl. Oxon.: 278. 1794, nom. inval. *Dicranum pulvinatum* Sw., Kongl. Vetensk. Acad. Nya Handl. 16: 240. 1795, nom. inval. *Fissidens pulvinatus* Hedw., Sp. Musc. Frond.: 158, tab. 40 fig. 1–3. 1801. *Fissidens pulvinatus* var. *communis* Hedw., Sp. Musc. Frond.: 158, nom.

illeg., incl. typ. spec. *Dicranum pulvinatum* (Hedw.) Lag., D. García & Clemente, Anales Ci. Nat. 5(14): 176. 1802. *Trichostomum pulvinatum* (Hedw.) F. Weber & D. Mohr, Bot. Taschenbuch: 109. 1807. *Campylopus pulvinatus* (Hedw.) Brid., Muscol. Recent. Suppl. 4: 75. 1818 ["1819"]. *Dryptodon pulvinatus* (Hedw.) Brid., Bryol. Univ. 1(1): 196. 1826. *Grimmia pulvinata* var. *vulgaris* Loeske, Laubm. Eur. Part I: 161. 1913, nom. illeg., incl. typ. spec. PROTOLOGUE: In tectis et muris vulgaris. TYPE: [no locality given] (lectotype, designated by Cao & Vitt (1986: 191), G!).

Fissidens pulvinatus var. *africanus* Hedw., Sp. Musc. Frond.: 159, tab. 40 figs. 4–6. 1801. *Dicranum africanum* (Hedw.) F. Weber & D. Mohr, Index Mus. Pl. Crypt.: 2. 1803. *Dicranum cribrosum* Brid., Muscol. Recent. Suppl. 1: 212. 1806 (NOTE: Erroneously considered legitimate by Muñoz (1999a: 168), this name is predated by *D. africanum*, see ICBN Art. 52.1). *Dicranum pulvinatum* var. *affine* Schleich. ex Brid., Muscol. Recent. Suppl. 4: 76. 1818 ["1819"], nom. inval., pro synon. *Dicranum pulvinatum* var. *africanum* (Hedw.) Schleich., Cat. Pl. Helv. ed. 4: 40. 1821. *Dryptodon cribrosus* (Brid.) Brid., Bryol. Univ. 1(1): 198. 1826. *Dryptodon obtusus* Brid., Bryol. Univ. 1(1): 198. 1826, nom. illeg. *Grimmia africana* (Hedw.) Arn., Mém. Soc. Linn. Paris 5: 243. 1827. *Grimmia pulvinata* var. *obtusa* Huebener, Muscol. Germ.: 194. 1833. "*Dryptodon obtusus* Hartm., Handb. Skand. Fl. ed. 3: 270. 1838" (NOTE: Error in Wijk et al. (1962: 176). Hartman neither combined nor described it as new). *Grimmia pulvinata* var. *africana* (Hedw.) Wilson, Bot. Antarct. Voy., Fl. Nov.-Zel. 2: 75. 1854. *Grimmia montana* var. *obtusa* Bruch & Schimp. ex Podp., Věstn. Klubu Přír. v Prostějově 7: 19. 1905 (NOTE: Error in Wijk et al. (1962: 393) for *Grimmia pulvinata* var. *obtusa* (Brid.) Huebener). *Grimmia obtusa* (Brid.) Loeske, Laubm. Eur. Part I: 165. 1913, nom. inval., pro synon. *Grimmia pulvinata* f. *obtusa* Vilh., Věstn. Král. Ceské Společn. Nauk, Tř. Mat.-Přír. 1924: 27. 1924, nom. illeg. PROTOLOGUE: [South Africa] Ad Cap. B. Spei lecta specimina mihi communicavit O. Swartz. TYPE: Ad Cap. bonae Spei lecta specimina misit O. Swartz (lectotype, designated by Muñoz (1999a: 168), G!).
Synonymized by Muñoz (1999a: 168).

Grimmia ecklonii Spreng., Syst. Veg. 4(2): 321. 1827. PROTOLOGUE: Ad saxa in monte leonis, C.B.S. Ecklon. TYPE: [South Africa] Swellendam. C.B.S. Ecklon (lectotype, here designated, BM!; isolectotype, BM!).

Dryptodon pulvinatus var. *crinitus* Bréb., Mousses de Normandie, Fasc. 5, nº 119. 1831, nom. nud. ORIGINAL MATERIAL and SPECIMEN: [France] Falaise. [L.A. de Brébisson] (FH!).

Grimmia cygnicolla Taylor, London J. Bot. 5: 45. 1846. PROTOLOGUE: [Australia] Swan River, Mr. James Drummond. TYPE: Swan River, N.H. James Drummond (lectotype, designated by Muñoz (1999a: 168), FH!; isolectotypes, FH!, H-SOL!, NY!).
Synonymized by Muñoz (1999a: 168).

Grimmia pygmaea Müll. Hal., Syn. Musc. Frond. 1: 787. 1849. *Grimmia trichophylla* var. *australis* Hampe ex Müll. Hal., Syn. Musc. Frond. 1: 788. 1849, nom. inval., pro synon. PROTOLOGUE: [Australia] Nova Hollandia, ad rupes graniticas: Preiss. TYPE: Grimmia trichophy.-australis Hampe. N. Holl. Juni loço (lectotype, here designated, H-SOL!; isolectotypes, H-BR!, PC!).

Grimmia callosa Müll. Hal. & Hampe, Linnaea 26: 498. 1855. PROTOLOGUE: [Australia] Australia felix, Barossa range.- Brown's hill [leg. F. Müller]. TYPE: Barossa-range. Australia felicis: Ferd. Müller (lectotype, designated by Muñoz (1999a: 168), JE!; isolectotype, MEL-1002648!). Synonymized by Muñoz (1999a: 168).

Guembelia obtusata Müll. Hal. & Hampe, Linnaea 26: 497. 1855. *Grimmia obtusata* (Müll. Hal. & Hampe) A. Jaeger, Ber. Tätigk. St. Gallischen Naturwiss. Ges. 1872–73: 70 (Gen. Sp. Musc. 1: 352). 1874. PROTOLOGUE: [Australia. Victoria] Australia felix [leg. F. Müller]. TYPE: Australia felix. [Leg. F. Müller] n° 7b (lectotype, here designated, MEL-1002652!).

NOTE: A duplicate of *Müller 7b* in NY! is *G. reflexidens* Müll. Hal.

Grimmia basaltica Mitt. ex Hook.f., Handb. N. Zeal. Fl.: 425. 1867. *Grimmia pulvinata* var. *basaltica* (Hook.f.) Dixon, New Zealand Inst. Bull. 3(4): 158. 1926, nom. illeg., non Vilh. 1924. PROTOLOGUE: [New Zealand] Middle Island: basalt rocks near Dunedin, Lindsay (Herb. Mitten). TYPE: Hills to south of the Forbury Heads near Dunedin, Ottago, New Zealand. November 19, 1861. Ex herb. Dr. Lauder Lindsay (lectotype, here designated, NY!; isolectotype, NY!).

Grimmia gibertii Mitt., J. Linn. Soc., Bot. 12: 100. 1869. PROTOLOGUE: [Uruguay] Brasilia, Monte Video, Gibert, n. 729, in Herb. Hooker. TYPE: n. 729. Montevideo, in terra. Augusto 1868. Gibert (lectotype, designated by Muñoz (1999a: 168), NY!; isolectotypes, FH!, H-BR!, NY!, PC!). Synonymized by Muñoz (1999a: 168).

Grimmia auresia Besch., Cat. Mouss. Alg.: 17. 1882. *Grimmia auresiaca* Besch. ex Kindb., Enum. Bryin. Exot., Suppl. 1: 91. 1889, nom. inval., orthogr. err. PROTOLOGUE: [Algeria] Aurès (Bal.[Balansa, B.]). TYPE: Aurès. Legit Balansa ex herb. Bescherelle (lectotype, designated by Muñoz (1999a: 168), PC!; isolectotype, F!). Synonymized by Muñoz (1999a: 168).

Grimmia hendersonii Renauld & Cardot, Rev. Bryol. 19: 86. 1892 ["Hendersoni"]. *Grimmia decipiens* var. *hendersonii* (Renauld & Cardot) Sayre, Bryologist 67(2): 163. 1964. PROTOLOGUE: [U.S.A. Oregon] Or. Leg. L.F. Henderson. TYPE: Rill Bed, Hood River [Co.], Oregon. March 31-89. L.F. Henderson n° 1933 (lectotype, here designated, PC!; isolectotypes, FH!, NY! 2 replicates).

Grimmia novae-zeelandiae R. Br. bis, Trans. & Proc. New Zealand Inst. 27: 414, pl. 31 fig. 16. 1895. PROTOLOGUE: [New Zealand] Damp rocks, Port Lyttelton hills; Mount Torlesse; Stewart Island; Dunedin; West Coast. [...] R.B. TYPE: Lyttleton Hills. R. Brown (lectotype, here designated, BM!).

Grimmia obovata R. Br. bis, Trans. & Proc. New Zealand Inst. 27: 416, pl. 32 fig. 23. 1895. PROTOLOGUE: [New Zealand] Rocks, Lyttelton hills; 1880. Collected by R.B. TYPE: (lectotype, here designated, fig. 23 in pl. 32, Brown, R., *op. cit.*). After an unsuccessful search for specimens referable to this taxon (BM, CHR, H), the illustration in the original publication is proposed as lectotype of the name. It unambiguously represents *Grimmia pulvinata* (Hedw.) Sm.

Grimmia pulvinata var. *asphaltita* Renauld & Cardot, Bull. Soc. Roy. Bot. Belgique 33: 118. 1895. PROTOLOGUE: [Israel] Hab. Judée: in deserto lacûs Asphaltitae inter Hebron et Zuweirat-el-Faga (Dr. G.E. Post, 1892. Herb. Boissier). TYPE: Plantae desertis Maris-Mortis. Entre Hebron et Zuweirat-el-Faga. Le 22 août 1892. Dr. E.G. Post (lectotype, here designated, PC!).

Grimmia pusilla R. Br. bis, Trans. & Proc. New Zealand Inst. 27: 417, pl. 33 fig. 26. 1895, nom. illeg., non F. Weber & D. Mohr 1803. PROTOLOGUE: [New Zealand] On rocks, Mount Self, near the Hurunui River March, 1893. Collected by R.B. TYPE: (lectotype, here designated, fig. 26 in pl. 33, Brown, R., *op. cit.*). After an unsuccessful search for specimens referable to this taxon (BM, CHR, H), the illustration in the original publication is proposed as lectotype of the name. It unambiguously represents *Grimmia pulvinata* (Hedw.) Sm.

Grimmia campbelliae Müll. Hal., Hedwigia 37: 162. 1898. PROTOLOGUE: Australia orientalis, Hume River: Miss Campbell 1881 in Hb. Melbourne. TYPE: Hume River. Miss Campbell. Baron Ferd. Mueller, Ph. & M.D. [ex MEL] (lectotype, here designated, JE!; isolectotypes, H-BR!, JE!). Synonymized by Muñoz (1999a: 168).

Grimmia tasmanica Müll. Hal., Hedwigia 37: 159. 1898. PROTOLOGUE: [Australia] Tasmania, sine loco natali: C. Stuart in Hb. Sonder 1840; in muris (wall of the catholic cemetery, Brisbane street): Weymouth 7.Oct.1890 in Hb. Dr. O. Burchard, qui 1891 pro determinatione misit. TYPE: ad muros: Weymouth 1890 (lectotype, here designated, PC-Besch!; isolectotypes, S!, W!).

Grimmia woollsiana Müll. Hal., Hedwigia 37: 162. 1898. PROTOLOGUE: Australia, New South Wales, Sydney: Rever. Dr. Woolls 1881 in Hb. Melbourne. TYPE: New South Wales; Rever. Dr. Woolls. Hb. C. Müller (lectotype, designated by Muñoz (1999a: 168), H-BR!; isolectotype, JE!). Synonymized by Muñoz (1999a: 168).

Grimmia leptotricha Müll. Hal., Hedwigia 38: 120. 1899. PROTOLOGUE: [South Africa] Prom. bonae spei, Somerset East, monte Boschberg, in rupibus: Prof. Mac Owan in Hb. Georg Winter 1883. TYPE: In monte Boschberg prope Somerset-East, Promont. bon. spei. leg. Mac Owan. [Rabenhorst-Winter, Bryotheca europaea et extraeuropaea, nº 1383] (lectotype, here designated, BM!).

Grimmia pulvinata var. *brevieri* Hérib., Mém. Acad. Sci. Arts Clermont-Ferrand, sér. 2, 14: 346. 1899. PROTOLOGUE: [France] Puy-de-Dôme.- Sur un mur à Ambert, tres rare (Brevière, 26 février 1895; fertile). TYPE: Puy-de-Dôme.- Vieux murs à Ambert, 26 février 1895. Leg. Brevière (lectotype, here designated, PC-Bonaparte!).

Grimmia fuscipila Müll. Hal. ex Paris, Index Bryol. Suppl.: 173. 1900, nom. nud. ORIGINAL MATERIAL: [India] As.: Himalaya sept. occid. (Dehra Doun). ORIGINAL SPECIMEN: Kumaun, Háwalbágh, 4–5000p., 26/4/1843. Leg. D. Jameson (H-BR! 2 replicates).

Grimmia subcurvula Kindb., Öfvers. Förh. Kongl. Svenska Vetensk.-Akad. 57(10): 1007. 1900. PROTOLOGUE: [Sweden] occurrit in Dalia prope Rostock in rupibus argillaceo-schistosis. TYPE: Dalsland, Rostock. 14/8 1898. N.C. Kindberg (lectotype, designated by Muñoz (1999a: 169), S!; isolectotype, PC!).
Synonymized by Muñoz (1999a: 169).

Grimmia hutchinsonii R. Br. bis, Trans. & Proc. New Zealand Inst. 35: 337, pl. 39, fig. 30. 1903. PROTOLOGUE: [New Zealand] On calcareous rocks near Weston, close to Oamaru. Collected by R. Brown, November, 1897. TYPE: Omaru 11/97 (lectotype, here designated, BM!).

Grimmia kaikouraensis R. Br. bis, Trans. & Proc. New Zealand Inst. 35: 337, pl. 40 fig. 31. 1903. PROTOLOGUE: [New Zealand] On damp rocks near Kaikoura. Collected by R. Brown. TYPE: Near Kaikoura, South I., New Zealand. On rocks. Coll. R. Brown ter. (lectotype, here designated, BM!).

NOTE: The material at BM is a mixture of *G. pulvinata* and *G. trichophylla*. Brown (1903: 337, pl. 40 fig. 31) used material of both species to describe and illustrate his new taxon. Some stems kept in a triangular packet have sporophytes and are gonioautoicous. They are selected as lectotype of the name. Dixon (1926: 157) considered *G. kaikouraensis* synonymous with *G. trichophylla*, although the specimen in his herbarium is annotated as identical to *G. pulvinata*. Recently, Greven (1998: 397) blindly followed Dixon's publication, although he annotated the BM specimen without realizing it was part of the type.

Grimmia pulvinata var. *sardoa* Herzog, Ber. Zürcherischen Bot. Ges. 15: 47. 1905. *Grimmia pulvinata* f. *sardoa* (Herzog) Laz., Žurn. Inst. Bot. Vseukrajins'k. Akad. Nauk 26–27: 206. 1938. PROTOLOGUE: [Italy. Sardinia] Auf Felsblöcken (Jurakalk) am Waldrand unter den Tonneri di Seui, am Weg zur Perdaliana, ca. 1100 m. 15. April 1904. TYPE: Sardinien. Auf Felsblöcken (Jurakalk) am Waldrand unter den Tonneri di Seui, am Weg zur Perdaliana, ca. 1100 m. 15. April, 04 (holotype, JE!).

Grimmia arachnoidea Dusén, Rep. Princeton Univ. Exped. Patagonia 8: 122. 1903, nom. nud. *Grimmia orbicularis* var. *patagonica* Cardot, Wiss. Erg. Schwed. Südpolar-Exp. 4(8): 284. 1908. PROTOLOGUE: [Chile] haute vallée du Rio Aysen ([Dusén] n° 587). TYPE: Patag. occ. in valle fluminis Aysen in saxis. Feb. 18 a. 1897. P. Dusén (lectotype, designated by Muñoz (1999a: 169), H-BR!; isolectotype, PC!).
Synonymized by Muñoz (1999a: 169).

Grimmia phyllorhizans Broth., Nat. Hist. Juan Fernandez 2(3): 419, tab. 26 fig. 21–24. 1924. PROTOLOGUE: [Chile] Masatierra: Villagra; ad saxa (129). TYPE: Juan Fernandez: Masatierra, steurösen v. om Villagra, st. 8. 7/1/1917. Carl o. Inga Skottsberg nr. 129 (lectotype, designated by Muñoz (1999a: 169–170), H-BR!; isolectotypes, FH!, NY!, UPS!).
Synonymized by Muñoz (1999a: 169).

Grimmia decurrentipilis Thér., in Felipp., Rev. Bryol. Lichénol. 2: 216, fig. 8. 1929. PROTOLOGUE: [Uruguay] La Paz, rochers; n° 837. TYPE: Herbario del Dr. F. Felippone. Plantas de la República del Uruguay. N° 837. Localidad: La Paz. Departamento: Canchones. Ad saxa. Marzo /913 (lectotype, designated by Muñoz (1999a: 170), PC!).
Synonymized by Muñoz (1999a: 170).

Grimmia montevidensis Thér., in Felipp., Rev. Bryol. Lichénol. 2: 216, fig. 7. 1929. PROTOLOGUE: [Uruguay] Montevideo, n° 839. TYPE: Herbario del Dr. F. Felippone. Plantas de la República del Uruguay. N° 839. Localidad: Montevideo. Departamento: Montevideo. Ad saxa. Marzo /913 (lectotype, designated by Muñoz (1999a: 170), BM!; isolectotype, PC!).
Synonymized by Muñoz (1999a: 170).

Grimmia pulvericola Müll. Hal. ex Bruehl, Rec. Bot. Surv. India 13(1): 123. 1931, nom. nud. ORIGINAL SPECIMEN: [India] NW Himalaya, Jaunsán 3000', Mai 1891. Leg. C.G. Rogers. Com. Levier (H-BR!).

Grimmia trichophylla var. *indianensis* Sayre, Bryologist 57: 21, fig. a–f. 1954. *Grimmia indianensis* (Sayre) Crum, Bryologist 68: 234. 1965. PROTOLOGUE: [U.S.A.] On sandstone along a small creek leading to Pine Creek on the east side of the road 2 miles southwest of Rainsville, Warren County, Indiana. Winona H. Welch 9694, 27 November 1937. Specimens in herbaria of DePauw University, New York Botanical Garden, Welch, Sayre. TYPE: On sandstone on small creek leading to Pine Creek on east side of road 2 m. southwest of Rainsville, Indiana, Warren County. 27 November 1937. Coll. W.H. Welch nº 9694. Det. G. Sayre (lectotype, here designated, NY!; isolectotypes, CANM-197537!, FH!, NY! 2 replicates).

Grimmia ahmadiana Nog., J. Hattori Bot. Lab. 16: 79. 1956. PROTOLOGUE: Pakistan: Swat State, Khaza Khela (S. Ahmad, ns. 102-typus, 108, Apr. 1954). TYPE: Pakistan: Khaza Khela, on the ground. Apr. 17, 1954, coll. S. Ahmad 102 (holotype, NICH-361549!).

DISTRIBUTION: ARG (*C.C. Hosseus 2220*, FH); ARM (*V. Manakyan 1841a*, ERE); AUS (*A.J. Taylor 6*, MEL); AUT (*Heimhold s.n.*, GOET); BEL (*C. Delogne s.n.*, FH); BRA (*D.M. Vital 9155*, FLAS); BGR (*S. Petrov s.n.*, MO); CAN (*J. Macoun s.n.*, FH); CHE (*O. Coulter s.n.*, TCD); CHL (*H. Deguchi 25159*, HIRO); CYP (*F. Unger s.n.*, JE); CZE (*M. Deyl s.n.*, MA); DEU (*M.R. Crosby 17700*, MO); DNK (*A. Hesselbo s.n.*, FH); DZA (*J.-P. Frahm 844526*, MA); ESP (*P. Seró B-356*, BCB); FIN (*J.O. Bomansson s.n.*, JE); FRA (*J.L. De Sloover 45145*, MA); GBR (*R.A. Pursell SVN-3*, MO); GEO (*P.H. Raven 27080*, MO); GRC (*V. Schiffner 157*, FH); HUN (*L. Vajda s.n.*, MA); IND (*W. Koelz 7544*, MO); IRL (*D.L. Kelly s.n.*, TCD); IRQ (*H.F.v. Handel-Mazzetti 1533*, FH); ISR (*P.H. Raven 22123*, MO); ITA (*M.R. Crosby 17699*, MO); LSO (*R.E. Magill 5863*, F); LBY (*Gimingham 146*, PC); MAR (*J. Muñoz s.n.*, MA); MKD (*T. Herzog 568*, JE); MEX (*E. Palmer 113*, MO); NOR (*N. Bryhn s.n.*, JE); NZL (*J. Child M2768*, F); POL (*K.G. Limpricht s.n.*, FH); PRT (*C. Sérgio et al. 31.3a*, BCB); SAU (*W. Frey & H. Kürschner 3133*, MA); SIB (*A.L. Abramova & I.I. Abramov s.n.*, MO); SUN (*Alexeenko 117-65*, LE); SVK (*J. Jedlička s.n.*, MO); SWE (*S.J. Lindgren s.n.*, FH); SYR (*H.K. Haussknecht 45*, JE); TKM (*U. Mamatkulov 2*, MO); TUR (*H.K. Haussknecht 67*, JE); TZK (*P. Polyakov s.n.*, NY); UKR (*A.S. Lazarenko s.n.*, NY); URY (*E.J. Gibert 729*, FH); USA (*A.T. Whittemore 3382*, MO); UZB (*L. Nazarenko 33*, LE); YUG (*M. Sabovljević s.n.*, MA); ZAF (*L. Rolfes 20*, JE).

59. Grimmia ramondii (Lam. & DC.) Margad., Lindbergia 1: 128. 1972. *Pterigynandrum ramondii* Lam. & DC., Fl. Franç. ed. 3 [original issue] 2: 462. 1805 ["ramondi"]. *Leucodon ramondii* (Lam. & DC.) Brid., Muscol. Recent. Suppl. 4: 135. 1818 ["1819"]. PROTOLOGUE: [France/Spain] dans les Pyrénées, par le C. Ramond. TYPE: Pyrenées. Herb. Cand. (lectotype, here designated, B!).

Bryum patens Dicks., Fasc. Pl. Crypt. Brit., Fasc. 2: 6, tab. 4 fig. 8. 1790, nom. inval. *Bryum patens* Hedw., Sp. Musc. Frond.: 86. 1801. *Pterigynandrum patens* (Hedw.) Brid., J. Bot. (Schrader) 1800(1(2)): 278. 1801. *Dicranum patens* (Hedw.) Sm., Fl. Brit. 3: 1213. 1804. *Trichostomum nudum* Schleich., Neues J. Bot. 1(2): 198. 1805, nom. nud. *Trichostomum patens* (Hedw.) D. Mohr, Ann. Bot. (König & Sims) 2: 545. 1806. *Fissidens patens* (Hedw.) Wahlenb., Fl. Lapp.: 334. 1812. *Campylopus patens* (Hedw.) Brid., Muscol. Recent. Suppl. 4: 73. 1818 ["1819"].

Dryptodon patens (Hedw.) Brid., Bryol. Univ. 1(1): 192. 1826. *Trichostomum patens* var. *majus* Hook. & Taylor, Muscol. Brit. ed. 2: 104. 1827, nom. illeg., incl. typ. spec. *Racomitrium patens* (Hedw.) Huebener, Muscol. Germ.: 198. 1833. *Grimmia arcuata* De Not., Mem. Reale Accad. Sci. Torino 39: 250. 1836, nom. illeg., incl. typ. spec. prior. *Grimmia patens* (Hedw.) Bruch & Schimp., Bryol. Europ. (fasc. 25–28) 3: 120, tab. 246. 1845, nom. illeg., non Hornsch. 1819. PROTOLOGUE: [United Kingdom] Ad rupes in rivulis alpinis Scotiae. Ben Nevis. Dicks. fasc. II. p. 6. t. IV. f. 8. TYPE: (BM? not seen).

Orthotrichum curvatum Brid., Bryol. Univ. 1: 791. 1827. *Grimmia curvata* (Brid.) De Sloover, Bull. Jard. Bot. État 37: 445. 1967. PROTOLOGUE: [Canada] In insula Terre Neuve habitat. TYPE: Terre Neuve. La Pylaie. [Parm, 1825, illegible] (holotype, B!).

Grimmia serrata Kindb., Rev. Bryol. 34: 90. 1907. PROTOLOGUE: Canada, Brit. Col., Rossland: J.M. Macoun 1902. TYPE: Canada, B.C., Old Glory mnt., Rossland. 15/8 02. J.M. Macoun (holotype, S!).
Synonymized by Allen (1995: 164).

Dryptodon patens var. *propagulifer* J.J. Amann, Fl. Mouss. Suisse 2: 140. 1918 ["propagulifera"]. PROTOLOGUE: [Switzerland] Valais: Emosson, 1750–1800 m., sur le gneiss (Amann) (BH 93.2.10) Graubünden: Albertitobel bei Davos-Platz, auf erratischen Gneiss, 1650–1700 m (Amann) (BH 93.2.8). TYPE: Graub. Albertitobel fur Davos. 24.5.90. Leg. Amann (BH 93.2.8) (lectotype, here designated, Z!; syntype ["Valais. Emosson March 1896. Leg. *Amann* (BH 93.2.10)"], Z!).

DISTRIBUTION: AUT (*K. Osterwald & V. Schiffner s.n.,* JE); BEL (*A. Lachmann s.n.,* PC); BGR (*L. Meinunger 14485,* JE); CAN (*J.M. Macoun s.n.*, S); CHE (*J. Amann s.n.*, Z); DNK (*C. Jensen s.n.*, NY); ESP (*P. Allorge s.n.*, FH); FIN (*R.W. Hartman s.n.*, H); FRA (*F.G. Kneiff & C.P.W. Maercker s.n.*, FH); GBR (*A.C. Crundell s.n.*, RM); GRL (*S. Berggren s.n.*, NY); IRL (*H.W. Lett s.n.*, TCD); ISL (*B. Jóhannsson s.n.*, ICEL); ITA (*A. Bottini s.n.*, FH); JPN (*T. Nagai 1531*, H-BR); NOR (*J. Heinrichs 202,* GOET); POL (*K.G. Limpricht s.n.*, FH); SUN (*Anonymous s.n.*, H-BR); SVK (*Z. Pilous s.n.,* JE); SWE (*J.E. Zetterstedt s.n.*, H); USA (*F.J. Hermann 24747*, RM); YUG (*M. Sabovljević s.n.*, MA).

60. Grimmia redunca Wilson ex Mitt., J. Linn. Soc., Bot. Suppl. 1: 43. 1859. *Grimmia redunca* Wilson, Hooker's J. Bot. Kew Gard. Misc. 9: 323. 1857, nom. nud. PROTOLOGUE: [India] In Himalayae orient. reg. alp., Sikkim, J.D. Hooker (No. 294). TYPE: 294. Jongri alt. 13.000. Dr. Hooker (holotype, NY!; isotypes, FH!, H-SOL!, LE!, PC!, PC-Besch!).

DISTRIBUTION: IND (*J.D. Hooker 294*, NY).

61. Grimmia reflexidens Müll. Hal., Syn. Musc. Frond. 1: 795. 1849. PROTOLOGUE: Chile australis inter Gr. consobrinam: Pöppig. TYPE: Inter *Dryptodon consobrinus* Kze. Chile. Leg. Pöppig (lectotype, designated by Muñoz (1998b: 394), BM!; isolectotypes, JE!, NY!, PC!).

Grimmia ramosa Schleich., Cat. Pl. Helv. ed. 3: 33. 1815, nom. nud.

Guembelia lamellosa Müll. Hal., Bot. Zeitung (Berlin) 12: 318. 1854. *Grimmia lamellosa* (Müll. Hal.) A. Jaeger, Ber. Tätigk. St. Gallischen Naturwiss. Ges. 1872/73: 72 (Gen. Sp. Musc. 1: 354). 1874. PROTOLOGUE: [France] In Pyrenaeis centralibus, in rupibus ad lacum Espingo, altitudine 5600 pedum, mense August. 1851 leg. Joh. Lange. TYPE: in Pyrenaeis centralibus, in rupibus ad lacum Espingo (lectotype, designated by Deguchi (1978: 207), PC!; isolectotypes, H-SOL!, JE!).
Synonymized by Muñoz (1998b: 394).

Grimmia sessitana De Not., Atti Reale Univ. Genova 1: 704. 1869. *Grimmia exannulata* Lindb. ex Broth., Acta Soc. Sci. Fenn. 19(12): 87. 1892, nom. inval., pro synon. *Grimmia alpestris* var. *sessitana* (De Not.) I. Hagen, Kongel. Norske Vidensk. Selsk. Skr. (Trondheim) 1909(5): 22. 1909. *Grimmia alpestris* f. *sessitana* (De Not.) Loeske, Laubm. Eur. Part I: 105. 1913. *Grimmia alpestris* var. *stomata* Loeske, Laubm. Eur. Part I: 104, figs. 26e,f,g, 27d,e,i,k. 1913, nom. illeg. PROTOLOGUE: [Italy] In praeruptis infra Hospitium della Valdobbia, in valle Sessitana, legit Carestia. TYPE: Frane alle scaturigini del Vogna, sotto l'ospizio della Valdobbia in Val Sessia. Carestia (lectotype, designated by Cao & Vitt (1986: 164), RO!; isolectotypes, BM! 2 replicates, FH!, PC-Thuret!, WRSL!).
Synonymized by Muñoz (1998b: 394).

Grimmia alpestris f. *hybrida* Chal., Pamietn. Fizyogr. 2: 68. 1882. *Grimmia alpestris* var. *hybrida* (Chal.) Chal., Enum. Musc. Frond. Tatr.: 56. 1886. *Grimmia sessitana* f. *longifolia* Loeske, Biblioth. Bot. 101: 116. 1930, nom. illeg., incl. typ. f. prior. PROTOLOGUE: [Poland and Slovakia, not explicitly stated]. TYPE: Tatra, Polnischer-Kaum, v.d. Felkaër-Thal. 27-VIII-1879. T. Chałubiński (lectotype, designated by Muñoz (1998b: 395), BP-36703!).
Synonymized by Muñoz (1998b: 395).

Grimmia anceps Boulay, Musc. France 1: 371. 1884. PROTOLOGUE: [France] Rochers siliceux du massif du Mt-Blanc, sur plusieurs points, vallée de Bérard, à gauche du torrent de l'Eau-Noire, en haut des Praz-d'Avaz, et ailleurs (Payot). TYPE: Massif du Mt. Blanc. Leg. Payot (lectotype, designated by Muñoz (1998b: 394), PC!; isolectotype, BP-36692!).
Synonymized by Muñoz (1998b: 394).

Grimmia tenerrima Renauld & Cardot, Bot. Gaz. 15: 40, pl. 6 A. 1890. PROTOLOGUE: [U.S.A.] Oregon: Mt. Hood, moist bluff towards the snow-line (L.F. Henderson). TYPE: Am. bor. Oregon, Mt. Hood, moist bluff above snow-line. Leg. Henderson, n° 1239 (lectotype, designated by Muñoz (1998b: 395), PC!; isolectotypes, FH!, NY! 2 replicates).
Synonymized by Muñoz (1998b: 395).

Grimmia grisea Cardot, Bull. Herb. Boissier, sér. 2, 6: 7. 1906. PROTOLOGUE: [South Georgia] Cumberland bay (Typus in herb. Cardot). TYPE: Géorgie du Sud: Cumberland bay. Leg. Skottsberg. 1902. n° 308 (lectotype, here designated, S!; syntypes ["*Skottsberg 307*"], H-BR!, S!).
Synonymized by Muñoz (1998b: 395).

NOTE: No specimens of *G. grisea* were found in PC.

Grimmia subpapillinervis Kindb., Rev. Bryol. 36: 98. 1909. PROTOLOGUE: [Canada] British Columbia 1908: Brinkman, com. Macoun. TYPE: on rocks, Ambean Valley, B.C. July 10, 1908. Coll. Brinkman (lectotype, designated by Muñoz (1998b: 395), CANM-198100!).
Synonymized by Muñoz (1998b: 395).

Grimmia subcaespiticia Schiffn., Ann. K. K. Naturhist. Hofmus. 27: 490. 1913. PROTOLOGUE: [Turkey] Ad rupes serpentinicas in regione montana Tauri Cataonici (Kurdistania occidentalis). Bekikara zwischen Kjachta und Malatja, am südlichen Talhang, 1600 m, c.sp., 18./VII. 1910 (Nr. 2422). TYPE: Kurdistania occidentalis: Taurus Cataonicus. In convallibus subalpinis prope vicum Bekikara inter urbem Malatja et vicum Kjachta, ad rupes. Substrato serpentinico; ca. 1600m. Leg. 18-VII-1910. Dr. Heinr. Frh. v. Handel-Mazzetti (Tageb.- No 502) (lectotype, designated by Muñoz (1998b: 395), FH!; isolectotype, H-BR!).
Synonymized by Muñoz (1998b: 395).

Grimmia asperitricha Dixon & Sainsbury, Trans. & Proc. Roy. Soc. New Zealand 75(2): 173. 1945. PROTOLOGUE: [New Zealand] On rocks near Hermitage, Mount Cook; coll. G.O.K.S. 29/1/1934, No. 573. On rock, De la Bêche, Tasman Glacier; coll. G.O.K.S. 20/1/1934, No. 756. On rock Malte Brun, Tasman Glacier; coll; G.O.K.S. 22/1/1934, Nos. 757, 771, 772. On rock, Pinnacles Ridge, Mount Ruapehu; coll. G.O.K.S. January, 1930, No. 799. TYPE: Tasman Glacier, near De La Beche Hut, South I., N.Z. Coll. G.O.K. Sainsbury n° 756, 20 Jan. 1934 (lectotype, designated by Muñoz (1998b: 395), BM!; syntypes ["*Sainsbury 753,*" "*Sainsbury 757,*" "*Sainsbury 771,*" "*Sainsbury 772,*" "*Sainsbury 799*"] F!, BM!).
Synonymized by Muñoz (1998b: 395).

Grimmia donniana f. *antarctica* Kuc, Rev. Bryol. Lichénol. 36: 659, figs. 1–2, 3 d–e, 4–5. 1969. PROTOLOGUE: Antarctic, "The Bunger Oasis", ad stationem nomine Dovrovolskii, leg. S. Rózycki, 25.VI.1959. Herb. Inst. Bot. Acad. Sc. Pol. Cracoviae. TYPE: Antarctica. Queen Mary Land, Bunger Hills: in the neighbourhood of the Polish Base. 25 June 1959. Coll. S. Różycki. Det. M. Kuc (holotype, KRAM!; isotype, MA-7386!).
Synonymized by Muñoz (1998b: 395).

DISTRIBUTION: AFG (*M. Köie s.n.*, ALTA); AND (*A. Sotiaux 3798*, BCB); ARG (*C. Skottsberg 976*, PC); ATA (*D.W.H. Walton 225*, MA); AUS (*L.G. Adams 1557*, FH); AUT (*V. Schiffner s.n.*, FH); BEL (*A. Cornet s.n.*, BP); BGR (*T. Simon s.n.*, BP); BOL (*M. Lewis 79-2092*, MA); CAN (*A. Brinkman 222*, MO); CHE (*J. Amann BH-48*, Z); CHL (*B. Sparre 4832*, FH); CHN (*C. Gao 22522*, ALTA); CZE (*I. Györffy s.n.*, JE); DEU (*O. Sendtner s.n.*, GLM); ESP (*J.L. Rupidera 2*, SALA); FRA (*J. Lange s.n.*, PC); GEO (*V.F. Brotherus 232*, H-SOL); ISL (*B. Jóhannsson 26913*, ICEL); IND (*W. Koelz 6774*, FH); IRN (*H.K. Haussknecht s.n.*, JE); ITA (*C.C. Townsend 92/547*, MA); JPN (*N. Takaki 133*, F); KAZ (*B. Allen 10554*, MO); KEN (*O. Hedberg 1923*, PC); MNG (*W. Schubert M196*, MO); NOR (*S.O. Lindberg s.n.*, H-SOL); NZL (*G.O.K. Sainsbury 753*, BM); PAK (*F. Duthie s.n.*, H); POL (*T. Chałubiński s.n.*, BP); ROM (*M. Péterfi s.n.*, BP); SFE (*B.C. Tan 97-199*, FH); SGS (*B.G. Bell 1188*, BM); SIB (*M. Ignatov 0/478*, MO); SUN (*T. Alexander s.n.*, H-BR); SVK (*L. Vajda s.n.*, BP); SWE (*O. Gjaerevoll s.n.*, TRH); TUR (*H.F.v. Handel-Mazzetti 2422*, FH); TZA (*Salt 46*, PC); UGA (*O. Hedberg 571*, MA); USA (*A. Nelson 2699*, RM); ZAF (*R.E. Magill 6122*, NY).

62. Grimmia teretinervis Limpr., Jahresber. Schles. Ges. Vaterl. Cult. 61: 216. 1884. *Schistidium teretinerve* Limpr., Jahresber. Schles. Ges. Vaterl. Cult. 61: 216. 1884, nom. alt. PROTOLOGUE: [Austria] Kalchstein bei Innervillgraten in Tirol vom Pfarrer Hieronymus Gander am 27. Juli 1882 gesammelt. TYPE: (BP? not found).

NOTE: Herbarium BP holds many duplicates of *G. teretinervis* collected by Gander in Innervillgraten in 1882 and identified subsequently by Limpricht, although not one matches exactly the date of the protologue. However, the publication shows both pagination and typographical errors, and under these circumstances a mistake on the transcription of the collection date could be the reason for the discrepancy.

Grimmia tenuis G. Roth, Eur. Laubm. 2: 691, tab. 62 fig. 4. 1905. *Grimmia barkeri* W. E. Nicholson & G. Roth, in G. Roth, Eur. Laubm. 2: 691, tab. 62. 1905, nom. inval., pro synon. PROTOLOGUE: [Switzerland] Bei Kandersteg in der Schweiz, dem Hotel Gemmi gegenüber, jenseits des Baches von Professor J. Barker im Herbst 1890 entdeckt. TYPE: Kandersteg river opposite the Hotel Gemmi. J. Barker. IX-1890 (lectotype, here designated, BM!; isolectotype, FH!).

DISTRIBUTION: AUT (*J. Baumgartner s.n.*, FH); CAN (*R.R. Ireland 15237*, FH); CHE (*C. Meylan BH-10*, Z); CZE (*I. Györrffy s.n.*, JE); DEU (*T. Herzog s.n.*, JE); HUN (*A. Boros s.n.*, BP); ITA (*J. Albrecht s.n.*, Z); ROM (*L. Vajda s.n.*, BP); SIB (*Zhuravlev 1734*, LE); SVN (*J. Breidler s.n.*, BP); SVK (*Z. Pilous s.n.*, BP); USA (*F.J. Hermann 20157*, G).

63. Grimmia tergestina Tomm. ex Bruch & Schimp., Bryol. Europ. (fasc. 25–28) 3: 126, tab. 258. 1845. *Grimmia tergestina* var. *eutergestina* Loeske, Laubm. Eur. Part I: 80. 1913, nom. inval. PROTOLOGUE: [Italy] In rupibus calcareis umbrosis loco "Bosco di Melara" prope Tergestum. TYPE: Bosco de Melara bei Triest [...] zwischen Longera und Bozzal. Ende März 1845. Tommasini (lectotype, designated by Cao & Vitt (1986: 178), BM!; isolectotypes, BM! 3 replicates, H-SOL!, PC!).

Grimmia kansuana Müll. Hal., Nuovo Giorn. Bot. Ital., n.s. 3: 109. 1896. PROTOLOGUE: China interior, prov. Kansu, Alpes ad captantes a flumine Tatung, 8000–9000 ped., in molis vetustis vulgaris, Augusto 1880: N.M. Przewalski. TYPE: Pl. a N.M. Przewalski collectae. China occidentalis. Regio Tangut (prov. Kansu) ad flum. Tatung, 8–9000'. 1/13 VIII.1880 (lectotype, here designated, PC!; isolectotype, PC!).

Grimmia philibertii Giacom., Atti Ist. Bot. Lab. Crittog. Univ. Pavia, ser. 5, 9: 200. 1950. *Grimmia orbicularis* × *G. tergestina* H. Philib., Ann. Sci. Nat. Bot., sér. 5, 17: 240, pl. 18 fig. 16–24. 1873, nom. inval., published as a formula. PROTOLOGUE: Aix, [...] chaine de collines appelée le Monteignez, [...] le Chicalon [vallée]. TYPE: Aix en Provence. 10 avril 1873. H. Philibert (lectotype, here designated, NY!).

Grimmia cucullata D.M. Hend., Notes Roy. Bot. Gard. Edinburgh 12: 614. 1958. PROTOLOGUE and TYPE: [Turkey] Prov. Ankara: above Tuz Gölu 25 km. north of Kochisar, c. 925 m., on bare rock and faces up to 45°, 29 July 1956, McNeill 342 (holotype, E!).

DISTRIBUTION: ARM (*V. Manakyan 2823a*, ERE); AUT (*D. Sabransky s.n.*, BP); BOL (*M. Lewis 79-243*, F); CHE (*J. Schleicher s.n.*, GOET); CHN (*B.C. Tan 93-1220*, FH); DEU (*L. Meinunger 1189*, JE); DOM (*W.R. Buck 8575*, NY); ESP (*J. Muñoz s.n.*, MA); FRA (*H. Philibert s.n.*, NY); GEO (*V.F. Brotherus 222*, H-SOL); HUN (*A. Degen s.n.*, BP); IRN (*P. Goldblatt 821*, MO); IRQ (*H.F.v. Handel-Mazzetti 1397*, FH); ITA (*E. Levier 22*, FH); MKD (*T. Herzog 678*, JE); PER (*P. Hegewald & E. Hegewald 5848A*, MO); ROM (*A. Degen s.n.*, BP); SAU (*W. Frey & H. Kürschner 1467*, MA); SFE (*V. Komarov 77*, LE); SIB (*N. Martinoff s.n.*, H-BR); SVK (*I. Györffy s.n.*, JE); TKM (*U. Mamatkulov 137*, MO); TUR (*J. McNeill 342*, E); UKR (*A.S. Lazarenko s.n.*, MA); USA (*D. Toren s.n.*, MA); YEM (*H. Kürschner 6201*, MA); YUG (*Loitlesberger s.n.*, JE).

64. Grimmia torquata Drumm., Musci Scotici, Vol. 2, nº 28. 1825. *Dryptodon torquatus* (Drumm.) Brid., Bryol. Univ. 1: 772. 1827. *Grimmia uncinata* subsp. *torquata* (Drumm.) Hampe, Flora 20: 285. 1837. *Zygodon torquatus* (Drumm.) Liebm. ex Beilschm., Jahresber. König. Schwed. Akad. Wiss. Fortschr. Bot. 1839–42: 430. 1845. PROTOLOGUE and TYPE: [United Kingdom] Clova mountains. [T. Drummond] (lectotype, designated by Geissler & Maier (1995: 505), E; isolectotypes, FH! 2 replicates, G!).

Grimmia cernua f. *atra* Nees & Hornsch., Bryol. Germ. 2(1): 178, tab. XXIV fig. 23b. 1827. PROTOLOGUE: [Austria] Ratiteberg 1816 aufgefunden (Mielichhofer! Hsch.! Funck! Laurer!); später auch auf Zirknitz-Alpe bei Döllach (Laurer!), und auf andern hohen Alpen von Ober-Kärnthen (Funck! Laurer!). TYPE: Herb. Nees v Esenb. Hernn Sturm (lectotype, here designated, G!).

Grimmia torta Nees & Hornsch., Bryol. Germ. 2(1): 179, tab. XXIV fig. 24. 1827. *Grimmia spiralis* var. *torta* (Nees & Hornsch.) Spruce, Musci Pyrenaici, nº 272. 1847. PROTOLOGUE: [Austria] Wächst auf Felsblöcken der höchsten Alpen, auf dem Rathhausberg in der Gastein (Laurer! Funck! Hrnsch.), auf der Gösnitz-Alpe (Hrsch.), auf dem Nafsfelder-Tauern (Funck! Laurer!), in der Nähe des Rauriser Tauernhauses (Laurer!), in den Schweizer Alpen (Fr. Nees v. Esenb.). TYPE: Herb. Nees v Esenb. ex alpibus (lectotype, here designated, G!; syntype ["Rauriser-Tauernhauses. *Laurer*"], BM!).

Grimmia streptophylla Kindb., Bot. Not. 1882: 187. 1882, nom. illeg., non Sull. 1859. *Grimmia tortifolia* Kindb., Christiania Vidensk.-Selsk. Forh. 1888(6): 19. 1888, nom. nov. *Grimmia stenophylla* Kindb. ex H. Möller, Ark. Bot. 26A(2): 67. 1934, nom. inval., pro synon. PROTOLOGUE: [Norway] Denna art år af mig funnen vid Kongsvold på Dovre ofvanför skogsregionen påomkring 950 meters höjd ofver hafvet. TYPE: Norveg., Dovre, prope Kongsvold, 900 metr. 23/7 1881 N.C. Kindberg = "*G. streptophylla*" specim. originalia! (lectotype, here designated, S!).

Grimmia prolifera Müll. Hal. & Kindb., in Macoun, Cat. Canad. Pl., Musci: 67. 1892. PROTOLOGUE: [Canada, locality not explicitly stated]. TYPE: Canada, B.C. Selkirk mts. 4300'. 29/7/1890. Leg. J. Macoun. Det. N.C. Kindberg (lectotype, here designated, S!; isolectotype, H-BR!).

Grimmia tortifolia subsp. *pellucida* Kindb., in Röll, Hedwigia 35: 64. 1896. *Grimmia torquata* subsp. *pellucida* (Kindb.) Paris, Index Bryol. ed. 2: 290. 1904. PROTOLOGUE: [U.S.A.] Yellowstone Nat. Park, Wyom., Upper Falls. No. 1478 [J. Röll]. TYPE: Yellowstone Nat. Park, Wyom., Upper Falls. 2/9 88. J. Röll 1478 (lectotype, here designated, S!).

Grimmia tortifolia subsp. *pseudotorquata* Kindb., Eur. N. Amer. Bryin. 2: 217. 1898. *Grimmia torquata* subsp. *pseudotorquata* (Kindb.) Paris, Index Bryol. ed. 2: 290. 1904. *Grimmia tortifolia* var. *pseudotorquata* (Kindb.) Paris, Index Bryol. ed. 2: 290. 1904, nom. inval., pro synon. PROTOLOGUE: Eur. Norway in Dovrefjeld and Gudbrandsdall. Sweden in Jemtland: P. Dusén. TYPE: [Sweden. Jamtlands:] Jemtl. Snasahogen 7/93. P. Dusén (lectotype, here designated, S!; syntypes ["Dovre, Kongsvold"] S! 9 replicates; ["Norway, Fåberg, Gudbrandsdall"] S!).

Racomitrium flettii Holz., Bryologist 7: 41, pl. 5. 1904 ["*Rhacomitrium*"]. *Grimmia flettii* (Holz.) Cardot, in Holz., Bryologist 8: 54. 1905. *Grimmia torquata* var. *flettii* (Holz.) Vaar., Bryologist

64: 324. 1962. PROTOLOGUE: [U.S.A.] collected by Prof. J.B. Flett on Mt. Tacoma (Mt. Rainier), at an altitude of 14,519 feet, near one of the steam jets issuing from a fissure in the side of the rim of the crater, Aug. 10, 1897. TYPE: crater of Mt. Tacoma (Rainier), alt. 14514 ft. J. B. Flett coll. Aug. 10, 1897 (lectotype, designated by Vaarama (1961: 324), ILL not seen; isolectotype, NY!).

Grimmia torquata f. *gracilis* Medelius ex H. Möller, Ark. Bot. 26A(2): 70. 1934. PROTOLOGUE: [Sweden] Härjedalen. Tännäs, Hamrafjället R. Oldberg; Jämtland. Are. Enafors 1819 A. Hülphers. TYPE: Härjedalen, Tännäs, Hammaren. R. Oldberg (lectotype, here designated, S!).

Grimmia torquata f. *longipila* H. Möller, Ark. Bot. 26A(2): 70. 1934. PROTOLOGUE: [Sweden. Norrbotten:] Lule lappmark, Kvikkjokk, Kadepakte 1867 H. Holmgren. TYPE: Lule lappmark, Kvikkjokk, Kadepakte. 27.7.1867. Leg. H. Holmgren (lectotype, here designated, S!).

Grimmia torquata f. *nigrescens* H. Möller, Ark. Bot. 26A(2): 70. 1934. PROTOLOGUE: [Sweden. Norrbotten:] Lule lappmark. Kvikkjokk. Njounjes 1921 M[ihi: H. Möller] m. fl. st. TYPE: Lule lappmark. Kvikkjokk. Njounjes. 7 Juli 1921. 1000 m. Leg. Hj. Möller (lectotype, here designated, S!).

DISTRIBUTION: AUT (*L. Molendo s.n.*, FH); BEL (*F. Gravet s.n.*, JE); BGR (*L. Meinunger 14398,* JE); CAN (*G.R. Brassard 10510*, FH); CHE (*H. Greven s.n.*, MA); CZE (*K.G. Limpricht s.n.*, FH); DEU (*L. Meinunger 1090.*, JE); ESP (*A.C. Crundwell 462*, NY); FIN (*V.F. Brotherus s.n.*, JE); FRA (*I. Thériot s.n.*, FH); GBR (*T. Drummond s.n.*, FH); GRL (*S.W. Bartlett H-5246*, NY); IRE (*D.L. Kelly s.n.*, TCD); ISL (*B. Jóhannsson 30100*, ICEL); ITA (*T. Herzog s.n.*, JE); NOR (*M. Haapasaari 8644*, JE); POL (*S. Lisowski s.n.*, PC); SFE (*O.M. Afonina s.n.*, NY); SUN (*O.A. Belkina & A.J. Likhachev s.n.*, MO); SWE (*C.F. Thedenius s.n.*, FH); USA (*F.A. MacFadden 16994*, MO).

65. Grimmia tortuosa Hook.f. & Wilson, London J. Bot. 3: 540. 1844. PROTOLOGUE: Falkland Islands. TYPE: Falkland Islands. Antarct. Exp. 1839–1843. J.D.H. (lectotype, designated by Muñoz (1999a: 174), BM!; isolectotypes, BM! 7 replicates, FH! 4 replicates, H-SOL!, NY!).

DISTRIBUTION: FLK (*J.D. Hooker 196*, BM).

66. Grimmia trichophylla Grev., Fl. Edin.: 235. 1824. *Grimmia affinis* Hornsch. ex Brid., Bryol. Univ. 1: 771. 1827, nom. inval., pro synon. *Dryptodon trichophyllus* (Grev.) Brid., Bryol. Univ. 1: 771. 1827. *Grimmia trichophylla* subsp. *eutrichophylla* Loeske, Laubm. Eur. Part I: 128, fig. 37, 38. 1913, nom. inval. PROTOLOGUE: [United Kingdom] On stone-walls about Edinburgh. On Arthur's Seat, and coast of Fife, Mr. Arnott. Lanar road, about seven miles from Edinburgh, and walls between Colinton and the Pentland Hills, abundant. [...] I discovered on a wall by the side of the road from Edinburgh to Lanark, about seven miles from the former place, in company with Mr Arnott. TYPE: near Edinburgh. Greville, 1821 (lectotype, designated by Geissler & Maier (1995: 505), E!; isolectotypes, FH!, H-SOL!).

Grimmia consobrina Kunze ex Müll. Hal., Syn. Musc. Frond. 1: 785. 1849. PROTOLOGUE: Chile australis: Pöppig. TYPE: Chile. Pöppig (lectotype, designated by Muñoz (1999a: 176), NY!; isolectotypes, BM! 2 replicates).
Synonymized by Deguchi (1984: 25).

Grimmia imberbis Kunze ex Müll. Hal., Syn. Musc. Frond. 1: 788. 1849. PROTOLOGUE: Chile australis, ad rupes: Pöppig. TYPE: Kze. Chile (lectotype, designated by Muñoz (1999a: 176), JE!).
Synonymized by Muñoz (1999a: 176).

Grimmia trichophylla var. *meridionalis* Müll. Hal., Syn. Musc. Frond. 1: 785. 1849. *Grimmia trichophylla* subsp. *meridionalis* (Müll. Hal.) Loeske, Biblioth. Bot. 101: 171. 1930. PROTOLOGUE: [Spain] Hispania, in colibus apricis pr. Granada et Escurial in Sierra de Guadarrama: W.P. Sch.–Aprili, Majo. TYPE: In coll. siccis prope Granatam. Schpr. lgt. (lectotype, designated by Muñoz (1999a: 176), H-SOL!; isolectotypes, JE!, NY!; syntype ["Escurial"] NY!).
Synonymized by Muñoz (1999a: 176).

Grimmia crispatula Müll. Hal. & Hampe, Linnaea 26: 499. 1855. *Grimmia crispata* Müll. Hal. & Hampe ex Mitt., Trans. Roy. Soc. Victoria 19: 55. 1882, nom. inval. PROTOLOGUE: [Australia. South Australia] Ad lapides rivuli Fift creek et Flinders range [leg. F. Müller]. TYPE: N. Holl. Flinders range (lectotype, designated by Muñoz (1999a: 176), H-SOL!; isolectotype, MEL-1002773; syntypes ["justa rivulum Fift creek"] H-SOL!, MEL-1002771!).
Synonymized by Muñoz (1999a: 176).

Grimmia subsquarrosa Wilson, J. Bot. 4: 231. 1866. *Grimmia muehlenbeckii* f. *propagulifera* Limpr., Laubm. Deutschl. 1: 764. 1889, nom. nov. *Grimmia ovata* var. *subsquarrosa* (Wilson) G. Roth, Eur. Laubm. 1: 414, tab. 32 fig. 10. 1903. *Grimmia trichophylla* f. *subsquarrosa* (Wilson) Loeske, Laubm. Eur. Part I: 130. 1913, nom. illeg., incl. typ. f. prior. *Grimmia trichophylla* subsp. *subsquarrosa* (Wilson) Dixon, Stud. Handb. Brit. Mosses ed. 3: 148. 1924. *Grimmia trichophylla* var. *subsquarrosa* (Wilson) A.J.E. Smith, J. Bryol. 9: 279. 1976. PROTOLOGUE: [United Kingdom] Hill of Moncrieff and Hill of Kinnoul, near Perth, May, 1865 [F.B. White]. TYPE: Hill of Moncrieff. May, 1865 (Dr. White) (lectotype, here designated, BM!; isolectotypes, BM! 3 replicates).

Grimmia stirtonii Schimp., Syn. Musc. Eur. ed. 2: 270. 1876 ["Stirtoni"]. *Grimmia trichophylla* f. *propagulifera* Limpr., Laubm. Deutschl. 1: 767. 1889. *Grimmia trichophylla* f. *stirtonii* (Schimp.) Loeske, Laubm. Eur. Part I: 129. 1913, nom. illeg., incl. typ. f. prior. *Grimmia trichophylla* subsp. *stirtonii* (Schimp.) Dixon, Stud. Handb. Brit. Mosses ed. 3: 147. 1924. *Grimmia trichophylla* var. *stirtonii* (Schimp.) H. Möller, Ark. Bot. 26A(2): 87. 1934. PROTOLOGUE: [United Kingdom] pr. Glasgow, ubi clar. Stirton 1866 in rupibus basalticis legit. TYPE: nr. Glasgow. June 1866. D. Stirton (lectotype, here designated, BM!).

Grimmia trichophylla f. *tenella* Besch., Cat. Mouss. Alg.: 18. 1882. PROTOLOGUE: [Algeria] Edough, rochers des caroubiers, mars 1861 (Let. [Letourneux, A.]). TYPE: Let. 18 mars 1861. Algérie (lectotype, here designated, PC!).

Grimmia watsonii Lesq. & James, Man.: 140. 1884 ["Watsoni"]. *Grimmia californica* subsp. *watsonii* (Lesq. & James) Kindb., Eur. N. Amer. Bryin. 2: 226. 1898. PROTOLOGUE: [U.S.A.] On rocks,

Dardanelles Cañon (Bolander); Alderney, Marin Co., California (Watson). TYPE: Alderney, Marin Co. Cal[ifornia]. Watson (lectotype, here designated, FH!).

Grimmia densa Kindb., Bull. Torrey Bot. Club 17: 271. 1890. *Grimmia depilata* Kindb., in Macoun, Cat. Canad. Pl., Musci: 69. 1892, nom. illeg., incl. typ. spec. prior. PROTOLOGUE: [Canada] On rocks, summit of Mount Benson, near Nanaimo, Vancouver Island, June 3, 1887. Alt. 3,000 feet. TYPE: N. Amer. Vancouver Island, Mt. Benson, 8/6 87. J. Macoun (lectotype, designated by Muñoz (1999a: 176), S!; isolectotype, CANM-198083!).
Synonymized by Muñoz (1999a: 176).

Grimmia cockaynei R. Br. bis, Trans. & Proc. New Zealand Inst. 27: 418, pl. 34 fig. 28. 1895. PROTOLOGUE: [New Zealand] Rocks, Mount Torlesse; January, 1886. Collected by R.B. TYPE: (lectotype, here designated, fig. 28 in pl. 34, Brown, R., *op. cit.*) After an unsuccessful search for specimens referable to this taxon (BM, CHR, H), the illustration in the original publication is proposed as lectotype of the name. It unambiguously represents *Grimmia trichophylla* Grev.

Grimmia finitima R. Br. bis, Trans. & Proc. New Zealand Inst. 27: 415, pl. 32 fig. 21. 1895. PROTOLOGUE: [New Zealand] Lyttelton hills; 1880. Collected by R.B. TYPE: (lectotype, here designated, fig. 21 in pl. 32, Brown, R., *op. cit.*) After an unsuccessful search for specimens referable to this taxon (BM, CHR, H), the illustration in the original publication is proposed as lectotype of the name. It unambiguously represents *Grimmia trichophylla* Grev.

Grimmia flexifolia R. Br. bis, Trans. & Proc. New Zealand Inst. 27: 417, pl. 33 fig. 25. 1895, nom. illeg., non Hampe 1860. *Grimmia brownii* Paris, Index Bryol. Suppl.: 172. 1900, nom. nov. *Grimmia flexuosa* Broth., Nat. Pflanzenfam. 3(1): 448. 1902, nom. inval., orthogr. err. PROTOLOGUE: [New Zealand] Rocks, Mount Torlesse; January, 1886. Collected by R.B. TYPE: Mt. Torlesse, North Canterbury, N.Z. R. Brown ter. (lectotype, here designated, BM!).

NOTE: The material in BM was annotated by Greven as *G. pulvinata* var. *africana*, although it is typical *G. trichophylla*. This author did not realize this specimen was a type of the name, and in his publication on New Zealand *Grimmia* (Greven, 1998: 397), he blindly followed Dixon (1926: 157) in considering *G. flexifolia* synonymous with *G. trichophylla*. The specimen was listed under *G. pulvinata* var. *africana*, however.

Grimmia petriei R. Br. bis, Trans. & Proc. New Zealand Inst. 27: 418, pl. 34 fig. 29. 1895. PROTOLOGUE: [New Zealand] Mount Torlesse; January, 1886. Collected by R.B. TYPE: Mount Torlesse 1/1/86 (lectotype, here designated, BM!).

Grimmia webbii R. Br. bis, Trans. & Proc. New Zealand Inst. 27: 419, pl. 34 fig. 31. 1895 ["webbi"]. PROTOLOGUE: [New Zealand] Palmer's Pass, head-waters of the River Conway; February, 1894. Collected by R.B. TYPE: On rocks, Palmer Pass 2/94 (lectotype, here designated, BM!).

NOTE: The type material at BM is a small quantity of pure *G. trichophylla*. It was annotated by Dixon as *G. trichophylla* var. *stirtonii*, although in his later publication he considered this name synonymous with *G. pulvinata* var. *obtusa*. Recently, Greven (1998: 396), after studying the type, and realizing it was a type, also considered it synonymous with *G. pulvinata* var. *africana*.

Grimmia sardoa var. *gracilis* Warnst. & M. Fleisch., Bot. Centralbl. 65(2/3): 299. 1896. *Grimmia trichophylla* var. *gracilis* (Warnst. & M. Fleisch.) Paris, Index Bryol. Suppl.: 176. 1900. *Grimmia trichophylla* f. *gracilis* (Warnst. & M. Fleisch.) Podp., Consp. Musc. Eur.: 288. 1954. PROTOLOGUE and TYPE: [Italy] Romano: Monti Albani, Mt. Cavo an Lavafelsen. 900 m. 7.6.1895 M. Fleischer [Fleischer & Warnstorf, Bryotheca europaea meridionalis. Cent. 1, nº 55. 1896] (lectotype, designated by Muñoz (1999a: 176), FH!; isolectotypes, GOET!, PC! 2 replicates).
Synonymized by Muñoz (1999a: 176).

Grimmia trichophylla var. *vivariensis* Husn., Musci galliae, fasc. 18, nº 861. 1897, nom. nud. in sched. ORIGINAL SPECIMEN: [France] Sur les pierres volcaniques à Labastide (Ardèche).- Mai 1892. Philibert (FH!).

NOTE: Erroneously listed in Wijk et al. (1962: 406) as *G. tergestina* var. *vivariensis*.

Grimmia trichophylla var. *teneriffae* Renauld & Cardot, Bull. Herb. Boissier, sér. 2, 2: 436. 1902. PROTOLOGUE: [Spain. Tenerife] Las Cañadas (nº 52) [A. Tullgren, 1896]. TYPE: Ténériffe: Las Cañadas. Leg. A. Tullgren, 1896 (lectotype, designated by Muñoz (1999a: 176), PC!).
Synonymized by Muñoz (1999a: 176).

Grimmia flexiseta Dusén, Rep. Princeton Univ. Exped. Patagonia 8: 122. 1903, nom. nud. ORIGINAL MATERIAL: Patagonia occ. ORIGINAL SPECIMEN: Argentina, ad lac. Nahualhuapi, ad saxa, 12/7/1897. Leg. P. Dusén (PC!).

Grimmia macropulvinata Dusén, Rep. Princeton Univ. Exped. Patagonia 8: 78, fig. 8, pl. 8 fig. 7. 1903. PROTOLOGUE: [Chile] Patagonia australis, ubi species verisimiliter in territorio fontinali fluminis Rio Chico in saxis lecta est. TYPE: Patagonia austr. Collected by J. B. Hatcher. Princeton Scientific Expedition to Patagonia 1896–97 (lectotype, designated by Muñoz (1999a: 176), S!; isolectotypes, FH!, NY!).
Synonymized by Deguchi (1984: 25).

Grimmia pycnophylla Dusén, Rep. Princeton Univ. Exped. Patagonia 8: 122. 1903, nom. nud. ORIGINAL MATERIAL: [Chile] Patagon. occ. ORIGINAL SPECIMEN: Patagonia occ. in valle fluminis Aysen in saxis. Jan. 30 a. 1897 P. Dusén (PC! 2 replicates, S! 3 replicates).

Racomitrium microphyllum M. Fleisch., Musc. Buitenzorg 1: 377. 1904. *Grimmia microphylla* (M. Fleisch.) M. Fleisch., Musc. Buitenzorg 4: 1650. 1923. PROTOLOGUE: [Indonesia] An Rinde. Ost-Java: Ardjoenogebirge, am Waliran! 2600 m. entdeckt (F.) [M. Fleischer, 1902]. TYPE: Ostjava. Ardjoenogeb. Waliran. 2700 m (holotype, FH!; isotype, FH!).
Synonymized by Deguchi (1986: 328).

Grimmia flexicaulis var. *dicksonii* Dusén, Bot. Not. 1905: 303. 1905. *Grimmia dicksonii* (Dusén) Dusén, Ark. Bot. 6(10): 14. 1907, nom. nov., nom. illeg., non Turner, 1804. PROTOLOGUE: [no locality given, but considered as: Patagonia occ., Río Aysén [Chile] in saxis. Febr.11.1897, leg. P. Dusén (cf . Deguchi, 1984: 25)]. TYPE: Patagonia, in vallis fluminis Aysén in rupibus. 11/2 1897. P. Dusén (lectotype, designated by Deguchi (1984: 25), S-PA; isolectotypes, FH! 2 replicates).
Synonymized by Deguchi (1984: 25).

Grimmia pachyphylla Dusén, Bot. Not. 1905: 302. 1905, nom. illeg., non Müll. Hal. 1885. PROTOLOGUE: [Chile] Patagonia occ. in valle flum. Aysen in saxis. TYPE: Patagonia occ. in valle flumini Aysen in rup. Febr. 11 a. 1897. P. Dusén (lectotype, designated by Muñoz (1999a: 176), H-BR!).
Synonymized by Muñoz (1999a: 176).

Grimmia apocarpa var. *sublaevipila* Cardot in Broeck, Bull. Jard. Bot. État 4: 256. 1914. PROTOLOGUE: [Belgium] Villers-la-Ville, mai 1882 (Delogne). Villécloye (Meuse) [...] à la côte de Thueux que M. J. Cardot [...] 5 janvier 1883. TYPE: Meuse: Villécloye, fissures de blocs rochers calcaires à la côte de Thueux. 5 Janvier 1883 (holotype, Blom (1996: 261), BR not seen; isotype, PC!).

Grimmia trichophylla var. *subincurva* H. Winter, Hedwigia 55: 101. 1914. PROTOLOGUE: [Spain. Tenerife] Levada do Gordon [...] [März–Mai 1912]. TYPE: Levada de Gordon (Monte) 900 mts. Mai 1912 leg. Dr. Winter (lectotype, designated by Muñoz (1999a: 176), JE!).
Synonymized by Muñoz (1999a: 176).

Grimmia trichophylla f. *viridis* Herzog, Kryptog. Forsch. 1(4): 281. 1919. PROTOLOGUE: [Former Yugoslav Rep. of Macedonia] Macedonien. TYPE: Fl. Mazedonien. An Felsen des Plateauberges bei Zovik, ca 650 m. Januar, 17 [1917]. Leg. Th. Herzog (lectotype, here designated, JE!).

Grimmia trichophylla f. *microtheca* Herzog, Bot. Jahrb. Syst. 60: 579. 1926. PROTOLOGUE: [Italy. Sardinia] Monte Santa Vittoria d'Esterzili bei etwa 1300 m. TYPE: Fl. v. Sardinien. Monte Santa Vittoria bei Esterzili. Zwischen 900 u. 1200 m. Mai, 06 [1906]. Leg. Th. Herzog (lectotype, here designated, JE!).

Grimmia gusindei J. Froehl., Ann. Naturhist. Mus. Wien 59: 117. 1953. PROTOLOGUE: [Chile] Fuegia: Canal Beagle, leg. M. Gusinde, II. 1922, no. 359. TYPE: Fuegia: Canal Beagle. Leg. M. Gusinde II.1922 n° 359. det. Froehl. 6.1.1953 (lectotype, designated by Muñoz (1999a: 176), S!).
Synonymized by Muñoz (1999a: 176).

DISTRIBUTION: ARG (*A. Donat 166*, JE); AUS (*A. Touw 18717*, FH); BEL (*F. Gravet s.n.*, JE); BOL (*M. Lewis 87-1812*, MO); CAN (*P.L. Redfearn 38635*, MO); CHL (*A. Hollermayer 72*, NY); COL (*S.P. Churchill 13329*, MO); CRI (*G. Dauphin 1690*, MO); CYP (*F. Unger s.n.*, H-SOL); DEU (*L. Meinunger 17190*, JE); DNK (*A. Hesselbo s.n.*, FH); DZA (*A. Letourneux s.n.*, PC); ECU (*E. Whymper 1347*, BM); ESP (*J.-L. De Sloover 4235*, BCB); FRA (*R.B. Pierrot 91007*, MA); GBR (*D.G. Long & B.M. Murray 14783*, MA); GRC (*V. Schiffner 11*, FH); GTM (*A.J. Sharp 2211*, TENN); HUN (*L. Marstaller s.n.*, JE); IDN (*Anonymous s.n.*, FH); IRL (*D.A. Webb s.n.*, TCD); ITA (*A. Berger s.n.*, NY); KEN (*S. Rojkowski 131*, F); LBN (*W.B. Evans 1034*, FH); MAR (*P. Davis & J. Davis D-49356*, NY); MEX (*Á. Cárdenas S. 5502*, MO); MKD (*T. Herzog 592*, JE); NZL (*D.H. Vitt 10666*, MO); PAN (*B. Allen 9093*, MO); PER (*P. Hegewald & E. Hegewald 5467*, MO); POL (*R. Ochyra s.n.*, MO); PRT (*A. Ervideira s.n.*, G); SAU (*P. König & H. Kürschner 1511*, MA); SWE (*O.J. Hasslow s.n.*, MO); SYR (*H.F.v. Handel-Mazzetti 1805*, FH); TUR (*J. Bornmüller 10085*, FH); USA (*A.T. Whittemore 3289*, MO); VEN (*D. Griffin III et al. 1328*, FLAS); YUG (*J. Baumgartner s.n.*, MO); ZAR (*J.L. De Sloover 13089*, NY).

67. Grimmia triformis Carestia & De Not., Comment. Soc. Crittog. Ital. 2: 102. 1866. *Grimmia donniana* var. *triformis* (Carestia & De Not.) Loeske, Laubm. Eur. Part I: 96, fig. 26d, 30. 1913. PROTOLOGUE: [Italy] Precipizii ombrosi del Tagliaferro in Val Sessia, Carestia. TYPE: In Tagliaferro, a quelche metro sotto la veta sul versante di Rima. 19 Ago 1863. Ab Carestia (lectotype, designated by Muñoz (1998b: 381), BM!; isolectotypes, BM! 2 replicates, BP-37541!, JE!, PC!).

Grimmia ganderi Limpr., Jahresber. Schles. Ges. Vaterl. Cult. 61: 215. 1884. *Grimmia donniana* var. *breviseta* Breidl. ex Loeske, Laubm. Eur. Part I: 95. 1913. PROTOLOGUE: [Austria] Wurde in Tirol bei Innervillgraten an glatten Schieferwänden am Wege nach "Kalchstein" bei 4000 Fuss vom Pfarrer Hieronymus Gander am 15. März 1884 gesammelt und als Grimmia triformis De Not. bestimmt, womit sie in der That am nächsten verwandt ist. TYPE: Tirol, Innervillgraten, an glatten Schieferwänden am Wege nach "Kalchstein" 4400' 15/3 1884 Gander (lectotype, designated by Muñoz (1998b: 381), BP-37547!; isolectotypes, JE! 2 replicates, H!).
Synonymized by Muñoz (1998b: 381).

DISTRIBUTION: AUT (*H. Gander s.n.*, G); CHE (*T. Herzog s.n.*, H); FRA (*R.B. Pierrot s.n.*, BCB); ITA (*A. Carestia s.n.*, JE); SIB (*M. Ignatov 0/474*, MA).

68. Grimmia trinervis R.S. Williams, Bull. New York Bot. Gard. 3(9): 124. 1903. *Coscinodon trinervis* (R.S. Williams) Broth., in Herzog, Biblioth. Bot. 87: 52. 1916. PROTOLOGUE: Juliaca, Peru, 3780 meters, on dry sandstone, May 15, 1902 (2814). TYPE: Peru. Juliaca. On dry sandstone. nº 2814. May 15 1902. Collected by R.S. Williams (holotype, NY!; isotype, F!).

DISTRIBUTION: BOL (*M. Lewis 79-1960*, F); PER (*H. Deguchi 29848*, HIRO).

69. Grimmia ungeri Jur., Ins. Cypern.: 169. 1865. *Grimmia intermedia* Braithw., J. Bot. 10: 198. 1872, nom. inval., pro synon. *Grimmia alpestris* subsp. *ungeri* (Jur.) Kindb., Bih. Kongl. Svenska Vetensk.-Akad. Handl. 7(9): 112. 1883. *Grimmia alpestris* var. *ungeri* (Jur.) Husn., Muscol. Gall.: 129. 1887. PROTOLOGUE: [Cyprus] Auf dem Troodos an Aphanit-Felsen. TYPE: In Olimpo Cypri. Vere 862. Leg. Dr. Fr. Unger (lectotype, designated by Muñoz (1998b: 398), BM!; isolectotypes, BM! 2 replicates, GOET! 2 replicates, H-SOL!, S!).

Grimmia brachyodon Austin, Bull. Torrey Bot. Club 6: 45. 1875. *Grimmia montana* var. *brachyodon* (Austin) Lesq. & James, Man.: 145. 1884. PROTOLOGUE: [U.S.A., locality not explicitly given]. TYPE: in rupibus schistosis montis Diablo Californiae (Bolander); etiam in alpibus Sierra Nevada montium (Brewer). [Sullivant & Lesquereux, Musci boreali-americani ed. 2, nº 215. 1866 ("1865")] (lectotype, designated by Muñoz (1998b: 398), FH!; isolectotypes, FH! 4 replicates, G!, H!, MO!, NY!, PC!).
Synonymized by Muñoz (1998b: 398).

Grimmia jamesii Austin, Bull. Torrey Bot. Club 6: 43. 1875. *Grimmia montana* var. *truncata* Lesq. & James, Man.: 145. 1884, nom. nov. PROTOLOGUE: [U.S.A.] *Grimmia orbicularis* James, in Watson's Catalogue of the Plants of the 40th Parallel, p. 403. [On rocks near Carson City, Nevada;

5,000 feet altitude. Watson (1,412)]. TYPE: Rocks. Carson. Nevada. Mai 26/68. nº 1412 (lectotype, designated by Muñoz (1998b: 398), FH!; isolectotypes, FH! 2 replicates).
Synonymized by Muñoz (1998b: 398).

Grimmia microtricha Müll. Hal. & Kindb., in Macoun, Cat. Canad. Pl., Musci: 70. 1892. PROTOLOGUE: [Canada] On rocks north of the railway close to Revelstoke, B.C., May 7th, 1890 (Macoun.). TYPE: N. Amer. Brit. Columbia, Revelstoke 7/5 1890 J. Macoun (lectotype, designated by Muñoz (1998b: 398), S!; isolectotype, CANM-198087!).
Synonymized by Muñoz (1998b: 398).

Grimmia montana var. *idahensis* Renauld & Cardot, Bot. Gaz. 30: 18. 1900. PROTOLOGUE: [U.S.A.] North Idaho: west end of Lake Pend d'Oreille (J.B. Leiberg, 1892. Ren. & Card., Musci Amer. sept. exsicc. no. 289). TYPE: Idaho: lac Pend d'Oreille. Leg. J.B. Leiberg, 1892. Renauld & Cardot, Musci Americae septentrionalis exsiccati no. 289 (lectotype, designated by Muñoz (1998b: 398), PC!; isolectotypes, CANM-197540!, FH! 2 replicates, NY!).
Synonymized by Muñoz (1998b: 398).

Grimmia pseudomontana Cardot & Thér., Bot. Gaz. 30: 18, pl. IV fig. 2. 1900. PROTOLOGUE: [U.S.A.] Idaho: near Moscow, on dry rocks (L.F. Henderson, 1894). TYPE: 4255. Dry rocks, near Moscow, Idaho. March 24/94 (lectotype, designated by Muñoz (1998b: 398), PC!).
Synonymized by Muñoz (1998b: 398).

Grimmia canadensis H. Winter, Hedwigia 55: 102, fig. 4. 1914, nom. illeg., non Kindb. 1897 ["cañadensis"]. *Grimmia ovalis* subsp. *canadensis* (H. Winter) Podp., Consp. Musc. Eur.: 279. 1954. PROTOLOGUE: [Spain] Teneriffa: Cañadas, in 2000, Höhe, auf Lavablöcken [März–Mai 1912]. TYPE: Teneriffa, Cañadas, 2000 mtr. auf Lavasblöcken. April 1912. Leg. Dr. Winter (lectotype, designated by Muñoz (1998b: 398), JE!; isolectotypes, JE! 5 replicates).
Synonymized by Muñoz (1998b: 398).

Grimmia mixta Herzog, Bot. Jahrb. Syst. 60: 573. 1926. PROTOLOGUE: [Italy. Sardinia] An Felsen der Ostseite der Punta La Marmora, etwa 1700 m, 2.Juni.1906. TYPE: An Felsen der Ostseite der Pta la Marmora, ca 1700m, 2.VI.1906 (holotype, JE!).

DISTRIBUTION: CAN (*J.F. Collins 3961*, FH); CYP (*F. Unger s.n.*, BM); ESP (*H. Winter s.n.*, JE); GBR (*J. Fergusson s.n.*, BM); ITA (*T. Herzog s.n.*, JE); MEX (*I.L. Wiggins 9177*, FH); USA (*F.A. MacFadden 17386*, FH).

70. Grimmia unicolor Hook., in Grev., Scott. Crypt. Fl. 3: 123. 1825. *Grimmia atrata* var. *minor* Drumm., Musci americani nº 57. 1828. *Dryptodon unicolor* (Hook.) Hartm., Handb. Skand. Fl. ed. 3: 272. 1838. PROTOLOGUE: [United Kingdom] Rocks near Bachnagairn, the Honourable D. Ogilvie's hunting lodge at the head of Clova; not rare, Mr. Drummond. TYPE: (lectotype, designated by Geissler & Maier (1995: 505), E not seen).

Grimmia norvegica Bryhn, Kongel. Norske Vidensk. Selsk. Skr. (Trondheim) 1899(3): 26. 1899. *Grimmia unicolor* var. *norvegica* (Bryhn) G. Roth, Eur. Laubm. 1: 413. 1903. *Grimmia unicolor* f. *norvegica* (Bryhn) Loeske, Laubm. Eur. Part I: 88. 1913. PROTOLOGUE: [Norway] In Bykle prope Bykleli ad rupes declivis juxta fluminem Otteraaen, altitudine supra mare 500 m., sociis

Dicrano longifolio et Racomitrio lanuginoso, pridie Nonas Augustas 1896 haud copiose lecta. TYPE: ad rupes (prope Bykleli) vallis Saetersdalen, Norwegia. 500m. Aug. 1896 leg. N. Bryhn (lectotype, here designated, S!).

DISTRIBUTION: AUT (*C.C. Townsend 92/530*, MA); CAN (*B. Allen 9568*, NY); CHE (*M. Jäggli s.n.*, GOET); CHN (*D.-K. Li 15197*, ALTA); ESP (*J.E. Zetterstedt s.n.*, H); ETH (*E.F. Gilbert 279*, NY); FIN (*V.F. Brotherus s.n.*, JE); GBR (*G.A.W. Arnott s.n.*, TCD); GRL (*W.C. Steere 62-1171*, NY); IND (*C.C. Townsend 89/441*, F); KAZ (*B. Allen 10562*, MO); MNG (*W. Schubert M152*, MO); NOR (*N.C. Kindberg s.n.*, S); SFE (*B.C. Tan 97-250*, FH); SIB (*M. Ignatov 0/464*, MA); SUN (*A.H. Brotherus & V.F. Brotherus 14*, H-SOL); SWE (*W.H. Arnell s.n.*, FH); USA (*F.J. Hermann 28754*, NY).

71. Grimmia wilsonii Greven, J. Bryol. 20: 398 fig. 1. 1998. PROTOLOGUE: New Zealand, Southern Alps, Mt. Cook National Park, summit of Sebastopol, 4820 ft., H.D. Wilson 3139, 18 November 1972, [holotype:] CHR 255583 (paratypes listed). TYPE: New Zealand, Southern Alps, Mt. Cook National Park, track to Hooker Glacier, *H.C. Greven NZ56* (paratype, MO!); *ibidem*, track to Sealy Tarn, *H.C. Greven NZ59* (paratype, BM!); New Zealand, Homer Saddle, *J. Child 4285* (paratypes, BM!, F!, MA!).

DISTRIBUTION: NZL (*H.C. Greven NZ56*, MO).

TAXA EXCLUDED FROM *GRIMMIA*

Amphidium tortuosum (Hornsch.) Cufodontis

Grimmia haliacalae Reichardt, Sitzungsber. Kaiserl. Akad. Wiss., Math.-Naturwiss. Cl., Abt. 1, 75: 567. 1877. PROTOLOGUE: [U.S.A. Hawaii] Maui; auf dem Gipfel und im Krater des Haliakala: Nr. 1900, 1902 [leg. Heinrich Wawra Ritter von Fernsee]. TYPE: Hawaische Inseln, Maui. Auf dem Gipfel und im Krater des Haliakala. N° 1900 Dr. Wawra (lectotype, here designated, W!; syntype ["*Wawra 1902*"], W!).

NOTE: The name was changed to *Grimmia haleakalae* in Wijk et al. (1962: 387), but according to the ICBN (Art. 60.1), the original spelling must be used.

Arctoa anderssonii Wich.

Grimmia cratericola Sakurai & Takaki, J. Hattori Bot. Lab. 6: 2, tab. 2. 1951. PROTOLOGUE: [Japan. Hauts sommets de la mte. Fuji] On soil in crevices of lava, altitude about 3700m. Coll. N. Takaki, July 22, 1950. No. 9068. TYPE: (holotype, private herbarium of N. Takaki, not seen; isotype, MAK not seen).
Synonymized by Deguchi (1978: 248).

Blindia acuta (Hedw.) Bruch & Schimp.

Grimmia gracilis Michx., Fl. Bor.-Amer. 2: 290. 1803. *Grimmia michauxii* Torr. ex Eaton, Man. Bot. ed. 3: 296. 1822. PROTOLOGUE: [U.S.A.] in Carolina; prior detexit D. Beauvois.
Synonymized by Allen (1995: 160).

Brachydontium olympicum (E. Britton) T.T. McIntosh & J.R. Spence, Bryologist 89: 200. 1986 [publ. 1987]. *Grimmia olympica* E. Britton, Bryologist 13: 59, pl. 7. 1910. PROTOLOGUE: [U.S.A.] Olympic mountains, Washington. Queets river valley, near Humes' glacier. Altitude 5,500 feet. T.C. Frye No. 614, Aug. 14, 1907. TYPE: (holotype, NY not seen).
Synonymized by McIntosh & Spence (1986).

Coscinodon sp.

Grimmia lawiana J.H. Willis, in Filson, The Lichens and Mosses of Mac.Robertson Land: 148, pl. 41. 1966. PROTOLOGUE: Ring Rocks, Mac.Robertson Land, Antarctica, leg. R. Filson No. 4225, 16 September 1962 (MEL). TYPE: Antarctica. MacRobertson Land, Ring Rock. 16 September 1973. Coll. R.B. Filson 4225 (holotype, MEL!; isotypes, MA!, KRAM!).

NOTE: *Grimmia lawiana* is the name used for sterile or otherwise poorly developed specimens collected in Antarctica. Some of the specimens so named are *G. plagiopodia*, but others, including the types, are not any *Grimmia* known to me. They have plicate leaves, plane hair-points and, in my opinion, should be referred to *Coscinodon* rather than to *Grimmia*.

Coscinodon arctolimnius (Steere) Steere, J. Hattori Bot. Lab. 42: 446. 1977 ["arctolimnia"]. *Grimmia arctolimnia* Steere, Bryologist 77: 230. 1974. PROTOLOGUE: Canada. Northwest Territories. District of Mackenzie: [...] Labine Point, near Eldorado Mine, Port Radium, McTavish Arm, Great Bear Lake (66°05'N Lat., 118°01'W Long.), 23 July 1948, Steere 10429. TYPE: Plants of the Region of Great Bear Lake, Northwestern Territoires, Canada. Labine Point, near Eldorado Mine, Port Radium, east end of McTarvish Arm. William C. Steere No. 10429. 23 July 1948 (holotype, NY!; isotypes, MO!, PC!).

Coscinodon bolivianus Broth., in Herzog, Biblioth. Bot. 87: 52. 1916. *Grimmia boliviana* (Broth.) F.J. Herm., Bryologist 79(2): 137. 1976. PROTOLOGUE: [Bolivia] Felsen am Huaillattanisee, alt. 4900 m, No. 2971. TYPE: Felsen am Huaillattanisee. 4900 m. Okt. 1911 leg. T. Herzog 2971 (lectotype, here designated, H-BR!; isolectotypes, H-BR!, PC!).

Coscinodon cribrosus (Hedw.) Spruce

Grimmia sinensianodon Müll. Hal., Nuovo Giorn. Bot. Ital., n.s. 5: 188. 1898. PROTOLOGUE: China interior, prov. Schen-si Sept., in loco dicto "Zu-lu" [Sao-y-san], Oct. 1896 [J. Giraldi]. TYPE: China interior, provincia Schen-si septentr. in loco dicto Zu-lu. 27 Oct. 1896 legit Rev. Jos. Giraldi, determ. C. Müll. nº 2063 (lectotype, designated by Cao & Vitt (1986: 167), H not seen; isolectotypes, PC! 2 replicates)

Synonymized with *G. caespiticia* by Cao & Vitt (1986: 167), but see Muñoz (1998b: 389).

Coscinodon humilis Milde

Grimmia brachyphylla Cardot, Bull. Herb. Boissier, sér. 2, 8(5): 333. 1908. PROTOLOGUE: [South Korea] Corée: Pomasa, rochers à 800 m (n. 218); No-in-tchi, rochers humides à 600 m (n. 590). TYPE: Corée: Pomasa. 800 m. Leg. Faurie, 1906. nº 218 (lectotype, designated by Muñoz (1998b: 391), PC!; isolectotypes, H!, H-BR!, NY!).

Synonymized with *G. montana* by Cao & Vitt (1986: 161), but see Muñoz (1998b: 391).

Coscinodon humilis Horik. & Nog., J. Jap. Bot. 14(1): 25–32. 1938, nom. inval. descr. jap. ORIGINAL MATERIAL: [Japan] Kyûsyû: Mt. Kudyû (ca 1500 m), prov. Bungo (A. Noguchi, no. 3298-typus, in Herb. Hiros. Univ., Nov. 1932). ORIGINAL SPECIMEN: (HIRO not seen).

Didymodon sp.

Grimmia velutina Kindb., Rev. Bryol. 23: 18. 1896. PROTOLOGUE: On dry rocks. Amer. Canada, Brit. Columbia: Macoun. TYPE: On the right bank of the Thompson River at Lytton. B.C. April 17th, 1889. Coll. Macoun (lectotype, here designated, CANM-198101!).

Didymodon rigidulus Hedw.

Campylopus curvifolius Brid., Muscol. Recent. Suppl. 4: 78. 1818 ["1819"]. *Dryptodon curvifolius* (Brid.) Brid., Bryol. Univ. 1(1): 201. 1826. PROTOLOGUE: [Poland] E Silesia sub nomine *Grimmia geniculata* missum accepi, Seliger invenit. TYPE: E comitatu Glatziam. Seliger. 1802 (lectotype, here designated, B!).

Jaffueliobryum arsenei (Thér.) Thér., Rev. Bryol. Lichénol. n. s. 1: 194. 1928. *Coscinodon arsenei* Thér., Smithsonian Misc. Collect. 81(1): 4 fig. 2. 1928. *Grimmia churchilliana* Crum, Bryologist 89: 24. 1986. PROTOLOGUE: [Mexico] Querétaro: Júrica, upon stones (11001). TYPE: Nº 11001. Mexico. Querétaro: Júrica. 1850 m. Leg. Fre. Arsène 5.2.1914 (holotype, PC!).

Jaffueliobryum wrightii (Sull.) Thér.

Grimmia julacea R.S. Williams, Bull. New York Bot. Gard. 6(21): 260. 1910, nom. illeg., non F. Weber & D. Mohr 1803. *Grimmia williamsii* Deguchi, Studies on Cryptogams in Southern Peru: 33. 1987, nom. nov. PROTOLOGUE: Arequipa, Peru, 2400 meters, on dry, exposed rock, Aug. 8, 1901 (2805) [R.S. Williams]. TYPE: Arequipa, 2400 meters, on dry rock. R.S. Williams Nº 2805. Aug, 8, 1901 (holotype, NY!; isotypes, F!, JE! 2 replicates).

Racomitrium spp.
NOTE: Names under *Racomitrium spp.* are ordered alphabetically.

Grimmia cylindropyxis Müll. Hal., Hedwigia 37: 166. 1898. PROTOLOGUE: Australia, Victoria, Mt. William: Dr. Sullivan 1883 legit et misit. TYPE: Mt. William. March 1883 [D. Sullivan] 9 (lectotype, here designated, MEL-1002777!).

Grimmia minuta Müll. Hal., Bot. Jahrb. Syst. 5: 81. 1884. *Racomitrium minutum* (Müll. Hal.) Ochyra & Matteri, Fragm. Florist. Geobot. 41: 1009. 1996. PROTOLOGUE: Kerguelen. TYPE: (lectotype, designated by Ochyra & Matteri (1996: 1009), S not seen).

Grimmia ochracea Müll. Hal., Bot. Jahrb. Syst. 5: 81. 1884. *Racomitrium ochraceum* (Müll. Hal.) Ochyra & Matteri, Fragm. Florist. Geobot. 41: 1008. 1996. PROTOLOGUE: Kerguelen. TYPE: (lectotype, designated by Ochyra & Matteri (1996: 1008), BM not seen; isolectotypes, BM, LE, S, none seen).

Guembelia procumbens Mitt., in F. Muell., Fragm. 11, Suppl.: 114. 1880, nom. nud. *Grimmia procumbens* Mitt., Trans. Roy. Soc. Victoria 19: 56. 1882. PROTOLOGUE: [Australia] Gippsland, Snowy River, F.v.M. [Ferdinand von Müller]. TYPE: NY holds specimens from both localities, labelled "Gipps Land. Dr. F. Müller. 1855" and "Snowy River. Victoria. Australia. Dr. Müller. 1855" respectively (syntypes, NY!).

Racomitrium crispulum (Hook.f. & Wilson) Hook.f. & Wilson
Grimmia defoliata Müll. Hal., Bot. Jahrb. Syst. 5: 81. 1884. PROTOLOGUE: Kerguelen. TYPE: Kerguelen. Rotag Cove, Dr. Naumann 1874 (lectotype, designated by Ochyra & Matteri (1996: 1009), S not seen; isolectotype, PC!).

Racomitrium didymum (Mont.) Lorentz
Grimmia depressa Müll. Hal., Flora 68: 421. 1885. PROTOLOGUE: [Argentina] Fuegia, Staten Island, Port Cook, ad truncos putridos in sylvis, Martio 1882. [Argentina] Fuegia australis, Burnst Island, ad ramos Chilostrii amelloidis in sylvis, Majo 1882: Spegazzini. TYPE: [Argentina] Ad truncos putridos, Stat. Isl., P. Cook 3/82. *Spegazzini 97* (lectotype, designated by Matteri & Muñoz (2000: 62), LPS!; isolectotype, PC not seen, Ochyra, pers. comm; syntype ["Burnt Isl. *Spegazzini 36*"], LPS!).
Synonymized by Matteri & Muñoz (2000: 62).

Grimmia depressa var. *terrestris* Müll. Hal., Flora 68: 421. 1885. PROTOLOGUE: [Argentina] Beagle Channel, in sinu Fandagaia [sic], ad glaream orae maritimae, Majo 1882: Spegazzini. TYPE: [Argentina] Yendegaia, ad glareum litoris 5/82. N° 31 (lectotype, designated by Matteri & Muñoz (2000: 63), LPS!; isolectotype, PC, Ochyra, pers. comm.).
Synonymized by Matteri & Muñoz (2000: 62).

Racomitrium ptychophyllum (Mitt.) Mitt., Handb. N. Zeal. Fl.: 426. 1867. *Grimmia ptychophylla* Mitt., in Linds., Trans. Bot. Soc. Edinburgh 8: 280. 1866. PROTOLOGUE: Otago, New Zealand. TYPE: Otago. W.L. Lindsay (holotype, NY!).

Schistidium spp.

NOTE: Names under *Schistidium spp.* are ordered alphabetically.

Grimmia abscondita Cardot, Wiss. Erg. Schwed. Südpolar-Exp. 4(8): 107, fig. 19. 1908. *Schistidium absconditum* (Cardot) Ochyra, Fragm. Florist. Geobot. 43: 104. 1998. PROTOLOGUE: [Argentina] Patagonie australe: lac San Martin, Rio Fosiles, vers 1000 m (Dusén, nº 5944). TYPE: (PC not seen).

Grimmia angustissima P. de la Varde, Ark. Bot., n.s. 3: 148, fig. 9. 1955. PROTOLOGUE: [Uganda] Ru [Mt. Ruwenzori]: Bukurungu Valley, in the side-valley on the S.E. slope of Mt. Gessi, 4400 m, on steep rock (680k, type). TYPE: Uganda: Ruwenzori, Bukurungu Valley in the side valley on the S.E. slope of Mt. Gessi, on steep rock. 4400 m.s.m. 4.4.1948. Coll. Olov Hedberg 680k. Det. PV (holotype, PC!).

Grimmia antipodum Müll. Hal., Hedwigia 37: 165. 1898. *Schistidium antipodum* (Müll. Hal.) Ochyra, Fragm. Florist. Geobot. 43: 105. 1998. PROTOLOGUE: [Australia] Tasmania, ad saxa, Frocy's Gully, Novbr. 1890: W.A. Weymouth in Hb. Burchard, qui pro determinatione 1891 misit. TYPE: Australia. Tasmania Frocy's Gully, Burchard 256 (lectotype, designated by Bremer (1980), H-BR not seen).

Grimmia austroapocarpa Müll. Hal. ex Watts. & Whitel., Proc. Linn. Soc. New South Wales, ser. 2, 27: 82. 1902, nom. nud. ORIGINAL MATERIAL: [Australia, no locality given]. ORIGINAL SPECIMEN: Parramatta. J. Whitelegge Aug 28/85 (H-BR!).

Grimmia calycina Herzog, Beih. Bot. Centralbl. 27(2): 354. 1910. *Schistidium calycinum* (Herzog) Ochyra, Fragm. Florist. Geobot. 43: 105. 1998. PROTOLOGUE: Bolivia: Cacaltaya, 4800 m, leg. O. Buchtien, März, 08. TYPE: Bolivia. Cacaltaya, 1908. Buchtien (isotype, Bremer (1980: 95), S not seen).

Grimmia celata Cardot, Bull. Herb. Boissier, sér. 2, 6: 7. 1906. *Schistidium celatum* (Müll. Hal.) B.G. Bell, Brit. Antarct. Surv. Bull. 63: 86. 1984. PROTOLOGUE: [South Georgia] Cumberland bay: connecting valley between S. and W. fiords. TYPE: Géorgie du Sud: Baie Cumberland, vallée entre les fjords S et O. 21/5 1902 Carl Skottsberg. Det. J. Cardot (lectotype, here designated, PC!).

Grimmia chubutensis Cardot & Broth., Kongl. Svenska Vetenskapsakad. Handl. 63(10): 26, tab. 2 fig. 6. 1923. *Schistidium chubutense* (Cardot) Ochyra, Fragm. Florist. Geobot. 43: 105. 1998. PROTOLOGUE: [Argentina] Patagonia andina: Territ. Chubut, Valle 16. de Octubre, ad saxa: Futaleufú, Casa Rees. TYPE: 432. Patagonie: Territoire Chubut, Vallée 16 Octobre, pierres dans la forêt de *Libocedrus chilensis*, mélange au *Barbula flagellaris*. 6.11.1908. C. Skottsberg (lectotype, PC!; syntypes ["*Skottsberg 439*"], H-BR!, UPS!).

Grimmia coloradensis Austin, Bot. Gaz. 2: 109. 1877 ["coloradense"]. *Schistidium coloradense* (Austin) Ochyra, Fragm. Florist. Geobot. 43: 105. 1998. PROTOLOGUE: [U.S.A.] Colorado, 1876, Brandegee (Rau). TYPE: Colorado, Nov. 9/76, Brandegee. Herb. Eugene A. Rau (lectotype, here designated, NY!).

Grimmia cupularis Müll. Hal., Bot. Jahrb. Syst. 5: 80. 1884. *Schistidium cupulare* (Müll. Hal.) Ochyra, Fragm. Florist. Geobot. 43: 105. 1998. PROTOLOGUE: Kerguelen. TYPE: Kerguelen. XI 74. Naumann (isotypes, H-BR!, PC!).

Grimmia diversifolia Kindb., Rev. Bryol. 32: 36. 1905. PROTOLOGUE: [Canada] 2. Found by J. Macoun 1904 in British Columbia, 5200–7500 feet above the sea. TYPE: Canada. B.C., Pipestone pass, 7500 f. 5/7 04. Macoun (lectotype, designated by Bremer (1980), S!).

Grimmia donatii Herzog & Thér., Repert. Spec. Nov. Regni Veg. 41: 290, tab. 236 e–n. 1937. *Schistidium donatii* (Herzog & Thér.) Ochyra & Matteri, Fragm. Florist. Geobot. 41: 1008. 1996. PROTOLOGUE: [Chile] Territorio Aisen. Regio Lago San Martin. Ventisquero, L.3, ca. 1500 m, leg. A. Donat, 13.V.33 (no. 48). TYPE: A. Donat Plantae patagonicae n° 48. Terr. Aisen. Reg. L.S.M. Ventisquero L.3. Alt. ca. 1500 m. Dat. 13.II.33 det Thériot, X.36 (holotype, JE!; isotype, JE!).

Grimmia fallax Dusén, Rep. Princeton Univ. Exped. Patagonia 8: 77, fig. 7, pl. 8 figs. 5–6. 1903. *Schistidium fallax* (Dusén) Ochyra & Matteri, Fragm. Florist. Geobot. 41: 1008. 1996. PROTOLOGUE: [Chile] Patagonia australis, ubi verisimiliter in territorio fontinali fluminis Rio Chico in saxis lecta est. TYPE: Princeton Scientific Expedition to Patagonia 1896–97. South Patagonia. Collected by J.B. Hatcher (holotype, Ochyra & Matteri (1996: 1008), S not seen; isotype, NY!).

Grimmia hyalinocuspidata Müll. Hal., in Neumayer, Int. Polarforsch., Deut. Exped. 2: 315. 1890. *Racomitrium hyalinocuspidatum* (Müll. Hal.) Kindb., Enum. Bryin. Exot., Suppl. 2: 108. 1893. *Schistidium hyalinocuspidatum* (Müll. Hal.) B.G. Bell, Brit. Antarct. Surv. Bull. 63: 90. 1984. PROTOLOGUE: [South Georgia] Austro-Georgia, ad rupes des Köppenberges 19. Majo 1883; Südwest-Gletscherthal, 7. Majo 1883. TYPE: Austro-Georgia, ad rupes Köppenberg Dr. Will. 19.V.1883 (syntype, PC!).

Grimmia kiyoshii S. Okamura, J. Coll. Sci. Imp. Univ. Tokyo 38(4): 12, fig. 5. 1916. PROTOLOGUE: [Japan] Hondo: Prov. Idu, in monte Amagi (Leg. Kiyoshi Fujii! 18.VIII.1914). TYPE: (isotype, Deguchi (1978: 220), TNS not seen).
Synonymized with *S. gracile* by Deguchi (1978: 220).

Grimmia konoi Broth., in Kono, Bot. Mag. (Tokyo) 20: 79. 1906. PROTOLOGUE: [in Japanese]. TYPE: (lectotype, designated by Deguchi (1978: 220), H-BR not seen).
Synonymized with *S. gracile* by Deguchi (1978: 220).

Grimmia longicaulis Dixon, J. Bot. 76: 221. 1938. PROTOLOGUE and TYPE: Limuru, Kenya Colony, 5 Jan. 1908; coll. G. Balbo (322) (holotype, BM!).

Grimmia macleana Rehmann ex Kindb., Enum. Bryin. Exot., Suppl. 1: 91. 1889, nom. nud. ORIGINAL SPECIMEN: [South Africa] Terra Capensis Orientalis, Valley of Desolation. leg. Mac Lea. Rehmann, Musci austro-africani cont., n° 493 (PC!).

Grimmia macrotyla Cardot & Broth., Kongl. Svenska Vetenskapsakad. Handl. 63(10): 26, tab. 2 fig. 5. 1923. *Schistidium macrotylum* (Cardot & Broth.) Ochyra, Fragm. Florist. Geobot. 43: 106. 1998. PROTOLOGUE: [Argentina] Patagonia andina: Territ. Chubut, Meseta Chalía frequentissime, alt. 12–1300 m.s.m. (382) [Leg. C. Skottsberg nº 382]. TYPE: Patagonie: territ. Chubut, meseta Chalia. Leg. C. Skottsberg, 1908. nº 382 (lectotype, here designated, PC!; isolectotypes, H-BR!, UPS!).

Grimmia occulta Müll. Hal., in Neumayer, Int. Polarforsch., Deut. Exped. 2: 314. 1890. *Schistidium occultum* (Müll. Hal.) Ochyra & Matteri, Fragm. Florist. Geobot. 41: 1008. 1996. PROTOLOGUE: [South Georgia] Austro-Georgia, inter muscos, 6. Febr. 1883 cum fruct. vetustis et junioribusi. TYPE: Austr. Georgia. Drekkappel. Dr. Will 6 II 1883 (lectotype, designated by Ochyra & Matteri (1996: 1008), HBG not seen; isolectotypes, PC!).

Grimmia perichaetialis P. de la Varde, Ark. Bot., n.s. 3: 149, fig. 10. 1955. PROTOLOGUE: [Uganda] Ru [Mt. Ruwenzori]: Mijusi Valley, E. slope of Mt. Speke, 4550 m, on moraine (580d, type); id., W. slope of Mt. Gessi, 4450 m, on stone, close to a small snow field (487); Bujuku Valley, S.W. slope of Mt. Speke, 4500 m (705a). Ke [Mt. Kenya]: N.W. slope, Buguret Valley, 4450 m (1915). TYPE: Uganda: Ruwenzori, Mijusy Valley, Eastern slope of Mt. Speke, on moraine. Coll. Olov Hedberg 580d, det. PV (holotype, PC!; syntypes ["*Hedberg 487*"], PC!; ["*Hedberg 705a*"], PC!; ["*Hedberg 1915*"], PC!).

Grimmia perichaetialis var. *pilifera* Demaret, Bull. Soc. Roy. Bot. Belgique 88: 92. 1956. PROTOLOGUE: Kenya: Mont Kenya, au-dessus de la Vallée Teleki, alt. 4500 m, mars 1949, Salt. nº 26 (holotype, BR not seen).

Grimmia perplexa Thér., Revista Chilena Hist. Nat. 27: 11, lám. 2. 1923. *Schistidium perplexum* (Thér.) Ochyra, Fragm. Florist. Geobot. 43: 106. 1998. PROTOLOGUE: [Chile] La Ermita, las Condes, alt. 1,200 m (leg. Bertho et Jaffuel, oct. 1919); Cajón del Arrayán, alt. 1,200 m (leg. Bertho, nov. 1919). TYPE: Chili: La Ermita (Las Condes) 1200 m. Leg. M. Bertho, oct. 1919 (lectotype, here designated, FH!; isolectotype, FH!).

Grimmia pseudorivularis Kindb., Eur. N. Amer. Bryin. 2: 234. 1898. *Schistidium pseudorivulare* (Kindb.) Ochyra, Fragm. Florist. Geobot. 43: 106. 1998. PROTOLOGUE: [Canada] Amer. r. Can.: Macoun. TYPE: Canada. Vancouver isl. 22/5/93 Macoun 159 (lectotype, here designated, S!).

Grimmia rivulariopsis R.S. Williams, Bull. Torrey Bot. Club 43: 327, pl. 19. 1916. *Schistidium rivulariopsis* (R.S. Williams) Ochyra, Fragm. Florist. Geobot. 43: 106. 1998. PROTOLOGUE: [Bolivia] Ollantaytambo, about 3,000 m., on rock in stream bed, May, 1915, Cook & Gilbert 753; and on rock, same place and collectors, 561. TYPE: Peru. Ollantaytambo; altitude about 3000 meters. No. 753 O.F. Cook & G.B. Gilbert Collectors. May 15, 1915 (lectotype, designated by Deguchi (1987: 34), NY!; isolectotype, F!).
Synonymized with *S. rivulare* by Deguchi (1987: 34).

Grimmia saxatilis Mitt., J. Linn. Soc., Bot. 12: 97. 1869. *Schistidium saxatile* (Müll. Hal.) Ochyra, Fragm. Florist. Geobot. 43: 106. 1998. PROTOLOGUE: [Ecuador] Andes Quitenses, in monte Chimborazo ad rivulos, Spruce, n. 132. TYPE: Chimborazo in rivulis. Spruce. 132 (lectotype, designated by Bremer (1980: 93), NY!; isolectotypes, FH!, PC! 2 replicates).

Grimmia scabrifolia Broth., Bernice P. Bishop Mus. Bull. 40: 12, tab. 4 fig. 13. 1927. PROTOLOGUE: [U.S.A. Hawaii] East Maui: on ground (B. 236) [D.D. Baldwin, n° 236]. TYPE: Insulae Sandwich in insula Maui orient., ad terram, 8000 p. 1876. Legit D.D. Baldwin sub n° 236 (isotypes, H-BR!, NY!).
Synonymized with *S. strictum* by Cao & Churchill (1995: 511).

Grimmia serratomucrata Müll. Hal., Bot. Jahrb. Syst. 5: 80. 1884. *Schistidium serratomucronatum* (Müll. Hal.) Ochyra, Fragm. Florist. Geobot. 43: 107. 1998. PROTOLOGUE: Kerguelen. TYPE: Kerguelens. Dr. Neumann. Nov. 1874 (lectotype, here designated, PC!).

Grimmia sinensiapocarpa Müll. Hal., Nuovo Giorn. Bot. Ital., n.s. 5: 187. 1898. *Schistidium sinensiapocarpum* (Müll. Hal.) Ochyra, Fragm. Florist. Geobot. 43: 107. 1998. PROTOLOGUE: China interior, prov. Schen-si sept., prope locum "Zu-lu", Oct. 1896 c.fr. deoperculatis [J. Giraldi]. TYPE: China interior, provincia Schen-si sept., Zu-lu in valle Lao-y-san. 28 Oct. 1896 legit Rev. Jos. Giraldi, determ Prof. C. Müller sub. n° 2051 (isotypes, PC! 2 replicates).

Grimmia syntrichiacea Müll. Hal., in Neumayer, Int. Polarforsch., Deut. Exped. 2: 315. 1890. *Schistidium syntrichiaceum* (Müll. Hal.) B.G. Bell, Brit. Antarct. Surv. Bull. 63: 97. 1984. PROTOLOGUE: [South Georgia] Austro-Georgia, Felsblöcke des südlichen Ufers der Landzunge, inter *Blindiam pallidifoliam* n. sp. specimina perpauca, 13. Oct. 1882. TYPE: Austro Georgia. 13.X.1882 Dr. Will (isotype, PC!).

Grimmia urnulacea Müll. Hal., in Neumayer, Int. Polarforsch., Deut. Exped. 2: 313. 1890. *Schistidium urnulaceum* (Müll. Hal.) B.G. Bell, Brit. Antarct. Surv. Bull. 63: 90. 1984. PROTOLOGUE: [South Georgia] Austro-Georgia, ad rupes am Ausgange des Brockenthales, cum *Gümbelia immerso-leucophaea* consociata, 23. Januario 1883. TYPE: (lectotype, designated by Ochyra (1990: 135), HBG not seen; isolectotype, S not seen).

Guembelia cyathocarpa Hampe, Linnaea 37: 515. 1872. *Grimmia cyathocarpa* (Hampe) Mitt., Trans. Roy. Soc. Victoria 19: 55. 1882. PROTOLOGUE: [Australia] New South Wales, Blue Mountains, planta riparia videtur. TYPE: [Australia] New South Wales. Blue Mountains. F.v. Müller (lectotype, here designated, BM!).

Guembelia lorentziana Müll. Hal., Linnaea 43: 450. 1882. *Grimmia lorentzi* (Müll. Hal.) Kindb., Enum. Bryin. Exot., Suppl. 1: 91. 1889. *Grimmia lorentziana* (Müll. Hal.) Paris, Index Bryol. 2: 529. 1895. *Schistidium lorentzianum* (Müll. Hal.) Ochyra & Matteri, Fragm. Florist. Geobot. 41: 1008. 1996. PROTOLOGUE: Argentinia subtropica Tucumanensis, in Sierra de Aconquija prope Tafi et in via ab Tafi per Quebrada de Monteros, ineunte aprilis 1872, [...]. TYPE: Argentinia. Tafi. Leg. Lorentz (lectotype, designated by Ochyra & Matteri (1996: 1008), JE!; isolectotype, BM!).

Guembelia praemorsa Müll. Hal., Linnaea 43: 452. 1882 ["Gümbelia"]. *Grimmia praemorsa* (Müll. Hal.) Kindb., Enum. Bryin. Exot., Suppl. 1: 91. 1889. PROTOLOGUE: Argentinia Catamarcana, supra Yakutula in convallis altae Granadillas, in Vayas altas [sic] alpina, ineunte Februarii 1872, [...]. TYPE: Argentinia, in alpinis prope Catamarca, Yakutula, in Vayas altas, convallis altas Grandillas [sic]. Febr. 1872 leg P.G. Lorentz (isotypes, BM!, JE!).

Schistidium amblyophyllum (Müll. Hal.) Ochyra & Hertel, Polish Botanical Studies 1: 26. 1990. *Grimmia amblyophylla* Müll. Hal., Syn. Musc. Frond. 1: 779. 1849. PROTOLOGUE: Insula Eremitae et Kerguelen ad Cap. Horn: J.D. Hooker. TYPE: Hermite Island, Cap Horn. Antart. Exp. 1839–1843 nº 122. J.D.H. (lectotype, designated by Ochyra & Hertel (1990: 26), BM not seen; isolectotypes, NY! 2 replicates).

Grimmia stylostegia Müll. Hal., Bot. Jahrb. Syst. 5: 80. 1884. PROTOLOGUE: Kerguelen. TYPE: Kerguelen. Dr. Naumann (isotype, PC!).

Grimmia antarctici var. *pilifera* E.B. Bartram, Ann. Missouri Bot. Gard. 25: 722. 1938. PROTOLOGUE: [Antarctica] Mary Bird Land, Edsel Ford Range, Mt. Rea-Cooper, P. Siple, F.A. Wade, S. Corey & O.D. Stancliff 27. TYPE: Byrd Antarctic Expedition II. Mosses of Marie Bird Land, Antarctica. Locality: Mt. Rea. Collected by Paul A. Siple nº 27. Determined by: Edwin B. Bartram. On granite sand (holotype, FH!; isotype, NY!).

Schistidium atrichum (Müll. Hal. & Kindb.) W.A. Weber, Phytologia 33: 106. 1976. *Grimmia atricha* Müll. Hal. & Kindb., in Macoun, Cat. Canad. Pl., Musci: 65. 1892. PROTOLOGUE: [Canada] On the sloping face of broken rocks at the summit of the mountains back of Sproat, Columbia River, B.C., alt. 4,000 feet, June 24th, 1890 (Macoun). TYPE: On broken rocks, alt. 4,000 feet. Sproat Mountain, Columbia River, B.C. June 24th, 1890. Coll. J. Macoun (isotypes, CANM-198078!, FH!, NY!).

Schistidium chenii (S.H. Lin) T. Cao, C.H. Gao & J.C. Zhao, J. Hattori Bot. Lab. 71: 69. 1992. *Grimmia himalayana* P.C. Chen, Report on Scientific Expedition of Qomolangma Region: 227. 1962, nom. illeg., non Mitt., 1859. *Grimmia chenii* S.H. Lin, Biol. Bull. Dept. Biol. Coll. Sci. Tunghai Univ. 60: 747. 1984, nom. nov. PROTOLOGUE and TYPE: China: Tibet, Himalayan region, Mt. Jolmo Lungma. Principal Lung-pu glacier, alt. 5450 m., on wet and moist rocks, 1959.VII.9, *Wang 86* (holotype, PE!).

Schistidium chrysoneuron (Müll. Hal.) Ochyra, Cryptog. Bryol. Lichénol. 7: 59. 1986. *Grimmia chrysoneura* Müll. Hal., Bot. Jahrb. Syst. 5: 80. 1884. PROTOLOGUE: Kerguelen. TYPE: Kerguelen. XII 74. Naumann (isotypes, H-BR!, PC!).

Grimmia antarctici Cardot, Bull. Herb. Boissier, sér. 2, 6: 15. 1906. PROTOLOGUE: Shetland du Sud: île Nelson, Harmony Cove, Terre Louis-Philippe: débarquement a. Archipel de Graham: île Paulet. TYPE: Archipel de Graham: île Paulet. Leg. Skottsberg, 1903. nº 449 (syntypes, FH!, H-BR!, NY!).

Grimmia antarctici var. *percompacta* E.B. Bartram, Ann. Missouri Bot. Gard. 25: 723. 1938. *Grimmia pilifera* var. *percompacta* H.A. Crum, Bryologist (index 1–60): 88. 1959, nom. inval. PROTOLOGUE: [Antarctica] King Edward VII Land: Rockefeller Mts., Mt. Helen Washington, P. Siple, F.A. Wade, S. Corey & O.D. Stancliff 68a. TYPE: Bird Antarctic Expedition II. Mosses of King Edward VII Land, Antarctica. Locality: Mt. Helen Washington. Collected by: Paul A. Siple nº 68a. Determined by: Edwin B. Bartram (holotype, FH!; isotype, NY!).

Schistidium cribrodontium (Herzog) Ochyra, Acta Bot. Acad. Sci. Hung. 28: 373. 1982. *Grimmia cribrodontia* Herzog, Repert. Spec. Nov. Regni Veg. 41: 287, tab. 235. 1937. PROTOLOGUE: [Uganda] Africa centralis: Ruwenzori, alpine Stufe, 4650 m, leg. I. Hauman VII/VIII, 32 (no. 854). TYPE: Ruwenzori. Etage alpin, 4650 m. Legit L. Hauman, julliet-août 1932. n° 854 (holotype, JE!).

Schistidium falcatum (Hook.f. & Wilson) B. Bremer, Lindbergia 6: 110. 1980. *Grimmia falcata* Hook.f. & Wilson, London J. Bot. 3: 539. 1844. *Schistidium falcatum* (Hook.f. & Wilson) Zanten, Marion and Prince Edwards Islands 2: 193. 1971, nom. inval., basionym not cited. PROTOLOGUE: Kerguelen's Island. TYPE: Kerguelen's Land. Antarct. Exp. 1839–43. J.D.H. n° 245 (isotypes, F!, FH!).

Grimmia fasciculata Dusén, Bot. Not. 1905: 302. 1905, nom. illeg., non Brid. 1819. PROTOLOGUE: [Chile/Argentina] Patagonia septentrionalis in saxis. TYPE: (lectotype, designated by Ochyra & Bell (1984: 78), S not seen; isolectotype, UPS not seen).
Synonymized by Ochyra & Bell (1984: 78).

Schistidium flaccidum (De Not.) Ochyra

Grimmia aethiopica Müll. Hal., Bot. Zeitung (Berlin) 13: 765. 1855. PROTOLOGUE: [Ethiopia] Abyssinia, in rupibus montium Simensium, inde a 9000 pedibus supra mare usque ad 13000 pedes ascendens: W. Schimper 1838 6. Januar. legit. TYPE: Schimperi iter Abyssinicum. Sectio secunda. 465. Schistidium pulvinatum Brid. In rupibus montium Simensium, indea [sic] 9000 pedibus supra mare usque ad 13000 pedes d. 6. Jun. 1838 U.i. 1842 (lectotype, here designated, NY!).

Schistidium canariense H. Winter, Hedwigia 55: 99, fig. 3. 1914. *Grimmia winteri* Luisier, Brotéria, Sér. Bot. 17: 39. 1919, nom. illeg., incl. typ. spec. prior. PROTOLOGUE: [Spain] Aufgenommen in Teneriffa an den Abhängen der Cumbre gegen Villa de Orotava in 1800 m Höhe von Lavablöcken [März–Mai 1912]. TYPE: Teneriffa. Abhängen der Cumbre gegen Villa de Orotava. 1800 mtrs. März 1912. Leg. Winter (lectotype, here designated, JE!; syntype ["April 1912"], JE!).

Schistidium frisvollianum H.H. Blom

Grimmia elatior subsp. *rufesgens* Kindb., Rev. Bryol. 34: 90. 1907. PROTOLOGUE: Canada, Hudson Bay, Baffin land, on earth: A.P. Low 1904, com. J. Macoun. TYPE: Canada, Hudson bay, Baffin land. 15/8 04. A.P. Low (lectotype, designated by Blom (1996: 87), S!).
Synonymized by Blom (1996: 87).

Schistidium heterophyllum (Kindb.) T.T. McIntosh, Bryologist 93: 483. 1990. *Grimmia heterophylla* Kindb., in Macoun, Bull. Torrey Bot. Club 17: 271. 1890, nom. nud. *Grimmia heterophylla* Kindb., Cat. Canad. Pl., Musci: 64. 1892. PROTOLOGUE: [Canada] between Nicola and Thompson rivers at Spence's Bridge, B.C., alt. 3,500 feet, May 28th, 1889 (Macoun.). TYPE: Spence's Bridge, B.C. Coll. Macoun. May 28, 1889. [Macoun, Canadian cryptogams n° 29] (isotype, FH!).

Grimmia concinodontoides Kindb., Bull. Torrey Bot. Club 17: 271. 1890, nom. nud. *Grimmia chloroblasta* Kindb., in Macoun, Cat. Canad. Pl., Musci: 64. 1892. *Grimmia coscinodontoides* Kindb., in Macoun, Cat. Canad. Pl., Musci: 64. 1892, nom. inval. *Grimmia coscinodontea* Kindb. ex Paris, Index Bryol. Suppl.: 172. 1900, nom. inval. PROTOLOGUE: [Canada] on dry rocks on the mountain overlooking the Thompson River at Spence's Bridge, B.C., alt. 3,500 feet, May 28th, 1889 (Macoun). TYPE: on rocks above Spence's Bridge. B.C. Coll. Macoun. Date May 28th. 1889 (isotypes, CANM-198081!, NY!).

Schistidium lancifolium (Kindb.) H.H. Blom, Bryophyt. Biblioth. 49: 55. 1996. *Grimmia pseudorivularis* subsp. *lancifolia* Kindb., Eur. N. Amer. Bryin. 2: 234. 1898. PROTOLOGUE: [Canada] Amer. Can.: Macoun. TYPE: Canada, Ottawa, wet rocks 23/9 93 J. Macoun (lectotype, designated by Bremer (1980: 95), S not seen).

Schistidium liliputanum (Müll. Hal.) Deguchi, J. Sci. Hiroshima Univ., Ser. B, Div. 2, Bot. 16: 229. 1978. *Grimmia liliputana* Müll. Hal., Nuovo Giorn. Bot. Ital., n.s. 5: 188. 1898. PROTOLOGUE: China interior, prov. Schen-si sept., in loco dicto "Zu-lu", Oct. 1896, inter *Grimmiam sinensi-anodontem* vigens [J. Giraldi]. TYPE: China interior, provincia Schen-si septentr. Zulu in valle Lao-y-san. 27 Oct. 1896, legit Rev. Jos. Giraldi, determ. C. Müller n. 2063b (lectotype, designated by Deguchi (1978: 229), FI not seen; isolectotypes, PC! 2 replicates).

Grimmia otii Sakurai, in Oti, Bot. Mag. (Tokyo) 62: 144, fig. 4. 1949. PROTOLOGUE: [Japan] Sikoku: Prov. Iyo (Leg. K. Oti no. 1228 - Type Novemb. 1945). TYPE: (holotype, Deguchi (1978: 229), MAK not seen).
Synonymized by Deguchi (1978: 229).

Schistidium maritimum (Turner) Bruch & Schimp.

Grimmia kindbergii Holz., Bryologist 14: 32. 1911. PROTOLOGUE: [U.S.A.] Catala, Alaska. Collector: Mr. G.C. Britton [1904] [...] Type in the herbarium of Prof. J.B. Flett, and of the author. TYPE: Catala, Alaska. Coll. G.C. Britton, summer of 1904 (isotypes, H-BR!, NY!).
Synonymized by Deguchi (1978: 233).

Schistidium olivaceum (Herzog) Ochyra & J. Muñoz, J. Bryol. 22: 142. 2000. *Grimmia olivacea* Herzog, Biblioth. Bot. 87: 56, fig. 18g–m. 1916. PROTOLOGUE: [Bolivia] Im oberen Chocayatal an Felsen, ca. 4400 m, No. 3587. TYPE: Plantae in itinere secundo per Boliviam lectae. An Felsen im oberen Chocayatal. ca. 4400m. Leg. Th. Herzog nº 3587. Juli 1911 (holotype, JE!; isotype, PC!).

Schistidium readeri (Broth.) Ochyra & J. Muñoz, J. Bryol. 22: 142. 2000. *Grimmia atricha* Müll. Hal., Hedwigia 37: 162. 1898, nom. illeg., non Müll. Hal. & Kindb. 1892. *Grimmia australasica* Mitt. ex Watts & Whitel., Proc. Linn. Soc. New South Wales, ser. 2, 27: 82. 1902, nom. inval., pro synon. *Grimmia readeri* Broth., Nat. Pflanzenfam. 1(3): 448. 1902, nom. nov. *Grimmia atrichoides* Paris, Index Bryol. ed. 2: 266. 1904, nom. illeg., incl. typ. spec. prior. *Schistidium atrichum* (Müll. Hal.) Ochyra, Fragm. Florist. Geobot. 43: 105. 1998, nom. illeg., non (Müll. Hal. & Kindb.) W.A. Weber. PROTOLOGUE: Australia, Victoria, ad saxa prope Melbourne legit Fr. M. Reader. Aug. 1886 et misit 1891 ex Dimboola.

Schistidium robustum (Nees & Hornsch.) H.H. Blom
Grimmia homodictyon Dixon, Rev. Bryol. 28: 12. 1901. PROTOLOGUE: On a detached block of limestone, crags south of Inchnadamph, Sutherlandshire, Scotland; July, 19, 1899, W.E. Nicholson, E.S. Salmon and H.N. Dixon. TYPE: [United Kingdom] Inchadamph, Sutherland, 19 July 1899, Nicholson, Salmon & Dixon (holotype, S not seen).
Synonymized by Blom (1996: 149).

Schistidium scabripes (E.B. Bartram) Deguchi, Studies on Some Patagonian Species of Grimmiaceae (Musci, Bryophyta): 27. 1984. *Grimmia scabripes* E.B. Bartram, Farlowia 2: 310. 1946. PROTOLOGUE: Chile: Prov. de Nuble, Recinto, Las Trancas, ad truncos Nothofagi antarcticae, alt. 1150 m., No 1003 type. TYPE: Chile: Prov. de Nuble, Recinto, Las Trancas, ad truncos Nothofagi antarticae, alt. 1150 m. Apr. 14, 1929. H. Roivainen no. 1003 (holotype, FH!).

Schistidium squamatulum (Herzog) Ochyra & J. Muñoz, J. Bryol. 22: 142. 2000. *Grimmia squamatula* Herzog, Biblioth. Bot. 87: 56, fig. 18 a–f. 1916. PROTOLOGUE: [Bolivia] An Felsen der Yanakakabastion, ca. 4600 m, spärlich, No. 3825/a. TYPE: An Felsen des Yanakakabastion ca. 4600 m. leg. Th. Herzog Juli 1911 (lectotype, designated by Muñoz (1999a: 184), JE!).

Schistidium subconfertum (Broth.) Deguchi, J. Sci. Hiroshima Univ., Ser. B, Div. 2, Bot. 16: 240. 1979 ["1978"]. *Grimmia subconferta* Broth., Symb. Sin. 4: 45. 1929. PROTOLOGUE: [China] S.: An nassen Schieferfelsen der Hg. St. auf dem Gonschiga sw von Muli gegen Dschungdien, 4650 m, c. sp. 6. VIII. 1915 (7493). TYPE: Prov. Schschwan austro-occid.: In montis Gonschiga a monast. Muli ad septentr. pagi Yünnanensis Yungming versus Dschungdien siti reg. alp., ad rupes udas. Substr. schistac. ca. 4650 m. 6.VIII.1915 leg. Hand.-Maz. (holotype, Deguchi (1978: 240), H-BR not seen; isotype, PC!).

Schistidium tenerum (J.E. Zetterst.) Nyholm, Ill. Moss Fl. Fennoscandia II. Musci 6: 775. 1969. *Grimmia tenera* J.E. Zetterst., Kongl. Svenska Vetenskapsakad. Handl. 13(13): 17. 1876. *Schistidium tenerum* (J.E. Zetterst.) Nyholm, Illustrated Moss Flora of Fennoscandia II. Musci 2: 146. 1956, nom. inval., basionym not cited. PROTOLOGUE: [Sweden] Alten, ad pedem montis Konghavnsfjellet; Talvig, in montibus juxta Talvigfjord praecipue prope amnem Isbärelfven. TYPE: Finm. Alten, Talvigfjord. 25/7/1868. Norge. J.E. Zetterstedt (lectotype, designated by Cao & Churchill (1995: 511), S not seen; isolectotypes, FH!, NY!).

Grimmia ehlei Arnell, Ark. Bot. 13(2): 63, Taf. 2. 1913. PROTOLOGUE: [Russia] Sibiria, in valle fluminis Lena, Kumachsur, 70° 30 lat. sept., ubi anno 1898 ab H. Nilsson-Ehle lectum. TYPE: Sibiria, in valle flum. Lena, Kumachsur, 70°30' lat. bor. 18/7/1898. Leg. H. Nilsson-Ehle. Det. H. Wilh. Arnell (isotypes, H-BR!, NY!).
Synonymized by Savicz-Lyubitzkaya & Smirnova (1970: 473).

Schistidium yaulensis (Broth.) Ochyra & J. Muñoz, J. Bryol. 22: 142. 2000. *Grimmia yaulensis* Broth., Bot. Jahrb. Syst. 56(2), Beibl. 123: 11. 1920. PROTOLOGUE: [Peru] An der Lima-Oroya-Bahn, Hacienda Arana bei Yauli; Felsenporphyr; 4400 m (373 ex p.) [Leg. A. Weberbauer]. TYPE: 373 ex p. Peru, an der Lima-Oroya-Bahn, Hacienda Arana bei Yauli, Felsen-Porphyr, 4400m 18/1/1902. Leg. A. Weberbauer (lectotype, designated by Muñoz (1999a: 184), H-BR!; isolectotype, PC! 2 replicates).

Tortula caninervis (Mitt.) Broth.

Grimmia cucullata J.S. Lou & P.C. Wu, Acta Phytotax. Sin. 18: 121, tab. 2 figs. 1–7. 1980, nom. illeg., non D.M. Hend. 1958. PROTOLOGUE: [Tibet] Xizang: Baingoin, 5600 m. alt., 27 VI.1976, K.Y. Lang 1317 (Type in Herb., Institute of Botany, Academia Sinica). TYPE: (holotype, PE not seen).
Synonymized by Cao & Vitt (1986: 233).

Trichostomum arcticum Kaal.

Grimmia spitsbergensis Bizot & Thér., Bull. Sci. Bourgogne 5: 70. 1935. *Grimmia ovalis* subsp. *spitsbergensis* (Bizot & Thér.) Podp., Consp. Musc. Eur.: 279. 1954. PROTOLOGUE: [Norway] Spitsberg, King's Bay. Leg. E. Maire, 1931. TYPE: Spitsberg. Kings Bay. Leg. Em. Maire 21.7.1931 (lectotype, here designated, PC!; isolectotype, NY!).
Synonymized by Frisvoll (1978: 158).

Verrucidens insularis (Mitt.) Ochyra & Matteri, Fragm. Florist. Geobot. 41: 1007. 1996. *Grimmia insularis* Mitt., J. Linn. Soc., Bot. 15: 73. 1876. PROTOLOGUE: [Antarctica] Heard Island [H.N. Moseley]. TYPE: Heard Island. Antarctic. Challenger Expedn. 1874 (lectotype, here designated, NY!; isolectotype, NY!).
Synonymized by Ochyra & Matteri (1996: 1007).

NAMES WITH TYPE MATERIAL NOT SEEN

Campylopus ovalis var. *politus* Myrin, Kongl. Vetensk. Acad. Handl. 1831: 258. 1832. PROTOLOGUE: [Sweden] I Dalsland på Henriksholms södra udde, vid Öxnäs bruk o. fl. St.

Campylopus pulvinatus var. *tenuis* Wahlenb., Fl. Suec. 2: 748. 1826. *Grimmia trichophylla* var. *tenuis* (Wahlenb.) Wijk & Margad., Taxon 8: 106. 1959. PROTOLOGUE: [Sweden] ß in campestribus saxis regionum interiorum ad Upsaliam et Vermlandiam usque passim.

Dicranum arcuatum Schleich., Cat. Pl. Helv. ed. 2: 28. 1808, nom. nud. *Trichostomum arcuatum* DC., Fl. Franç. ed. 3, 6: 218. 1815. *Trichostomum patens* var. *arcuatum* (DC.) Duby, Bot. Gall. 2: 574. 1830. PROTOLOGUE: [Switzerland] [...] dans les Alpes, par M. Schleicher; [France] dans les Cévennes, par M. Grateloup.

NOTE: A syntype ["Schleicher. Originaux utilisés par A.-P. de Candolle pour la rédaction de la Flore française. *Trichostomum arcuatum* DC"] in G is *Grimmia elatior*. *Trichostomum arcuatum* predates the name *G. elatior* and should be used according to the ICBN. Since another syntype is listed in the protologue, there exists the possibility of typifying on another element to maintain nomenclatural stability and avoid the undesired name change. If the other syntype also proves to be *G. elatior*, a well-known and long-used name, in my opinion the latter name should be proposed for conservation to preserve nomenclatural stability.

Dicranum pulvinatum var. *argentatum* Turner, Musc. Hibern. Spic.: 78, tab. 3 fig. 1. 1804. *Dryptodon pulvinatus* var. *argentatus* (Turner) Brid., Bryol. Univ. 1(1): 197. 1826. PROTOLOGUE: [United Kingdom] in Eblanae muris a D. Scott, et in muris prope Powerscourt, a D. Stokes lecta.

Dicranum pulvinatum var. *muticum* Chevall., Fl. Gén. Env. Paris 2(1): 41. 1827. PROTOLOGUE: [France. Paris].

Dryptodon caespiticius var. *muticus* Brid., Bryol. Univ. 1(1): 201. 1826. *Grimmia alpestris* var. *mutica* (Brid.) De Not., Syllab. Musc.: 244. 1838. PROTOLOGUE: In eodem loco [Switzerland. In monte sancti Bernhardi Helvetiae].

Dryptodon hartmanii f. *rivularis* Loeske, Moosfl. Harz.: 194. 1903. *Grimmia hartmanii* f. *rivularis* (Loeske) Loeske, Biblioth. Bot. 101: 179. 1930. PROTOLOGUE: Am Bodeufer unterhalb Tresebg. [Germany. Treseburg].

Dryptodon hartmanii var. *alatus* Katić, Prosv. Glasn. 4: 885–886. 1907. PROTOLOGUE: [in Serbo-Croatan: Rocks at Deli Jovan, below Crni Vrh, at Jasenicka river source, ca. 1050 m. May. Leg. D. Katić].

Dryptodon hartmanii var. *giganteus* Jeanb. & Renauld ex Warnst., Krypt.-Fl. Brandenburg 1(2): 302. 1904. *Grimmia hartmanii* f. *gigantea* Jeanb. & Renauld, Guide Bryol. Pyrénées. Part 2: 55. 1885, nom. nud. *Grimmia hartmanii* var. *gigantea* (Warnst.) Podp., Consp. Musc. Eur.: 289. 1954. PROTOLOGUE: aus den Pyrenäen bei 1750 m Meereshöle.

Dryptodon ovatus var. *denticulatus* Brid., Bryol. Univ. 1(1): 203. 1826. PROTOLOGUE: In Britanniae rupibus. Green.

Dryptodon ovatus var. *elongatus* Brid., Bryol. Univ. 1(1): 203. 1826. PROTOLOGUE: [Italy] In Calabriae rupibus Clar. Reynier legit et communicavit.

Dryptodon patens f. *cataractarum* Vilh., Acta Bot. Bohem. 2: 52. 1923. PROTOLOGUE: [Slovakia] [...] montium tatrensium regionis alpinae (supra Velické pleso, 1900 m; supra žabí pleso, 2000 m; Mlinica sub lacum Skok, an. 1919 Vilh.).

Dryptodon patens f. *humilis* Vilh., Acta Bot. Bohem. 2: 52. 1923. PROTOLOGUE: [Slovakia] [...] regionis alpinae in Sudetis (Braunkessel apud Pecer, Pansčká louka in montibus Corconticis) et in Tatris (Mlinica circum lacum Skok, 1820 m; sub Rysy, 2100 m, sub Velické pleso in monte Gerlach, sub Zelené pleso apud Krivaň, an. 1919 Vilh.).

Dryptodon patens f. *pilifer* Vilh., Acta Bot. Bohem. 2: 53. 1923 ["pilifera"]. PROTOLOGUE: [Slovakia] [...] ca. 1600–2000 m in Tatris sub 5 ples in Malá studená dolina (an. 1919 Vilh.).

Dryptodon patens f. *robustus* Vilh., Acta Bot. Bohem. 2: 52. 1923 ["robusta"]. PROTOLOGUE: [Slovakia] [...] montium tatrensium (Mlinica ca. 1500 m, an. 1919 Vilh.).

Dryptodon pulvinatus var. *canus* Hartm., Handb. Skand. Fl. ed. 3: 270. 1838. *Grimmia pulvinata* var. *cana* (Hartm.) Hartm., Handb. Skand. Fl. ed. 5: 376. 1849. *Grimmia pulvinata* f. *cana* (Hartm.) Mönk., Laubm. Eur.: 365. 1927. PROTOLOGUE: ß *D. sudeticus* från Cimbritshamn [Norway], ed. 2.

Grimmia albida Spreng., Syst. Veg. 4(2): 321. 1827, nom. illeg., non F. Weber & D. Mohr 1803. *Grimmia albicans* Spreng. ex H. Möller, Ark. Bot. 26A(2): 135. 1934, nom. inval., orthogr. err. PROTOLOGUE: [South Africa: Caput Bon Spei] C.B.S. Ecklon.

Grimmia alpestris f. *longipila* Podp., Veröff. Geobot. Inst. Rübel Zürich 1: 245. 1924. PROTOLOGUE: [Switzerland] Zernez: [...] oberhalb der Buffalorawiese.

Grimmia alpestris var. *microstoma* Bruch & Schimp., Bryol. Europ. (fasc. 25–28) 3: 129. 1845. *Grimmia subsulcata* Limpr., Laubm. Deutschl. 1: 757. 1889. *Grimmia sessitana* var. *subsulcata* (Limpr.) Breidl., Mitt. Naturwiss. Vereines Steiermark 28: 88. 1891, nom. illeg., incl. typ. var. prior. *Grimmia alpestris* subsp. *subsulcata* (Limpr.) Kindb., Eur. N. Amer. Bryin. 2: 221. 1898. *Grimmia microstoma* (Bruch & Schimp.) Dalla Torre & Sarnth., Fl. Tirol 5: 260. 1903. *Grimmia alpestris* f. *subsulcata* (Limpr.) Loeske, Laubm. Eur. Part I: 104. 1913. *Grimmia alpestris* var. *subsulcata* (Limpr.) Broth., Nat. Pflanzenfam. ed. 2, 10: 309. 1924. *Grimmia sessitana* f. *subsulcata* (Limpr.) Loeske, Biblioth. Bot. 101: 117. 1930. PROTOLOGUE: Ad rupes siccas per Alpes editiores Helvetiae, Germaniae, Delphinatus et Norvegiae [no locality given for the variety].

NOTE: In a previous publication, Muñoz (1998b: 394) erroneously considered *Grimmia subsulcata* as having a different type from *Grimmia alpestris* var. *microstoma.*

Grimmia anodon var. *brevipila* C.E.O. Jensen, in Dusén, Bih. Kongl. Svenska Vetensk.-Akad. Handl. 27, Afd. 3(1): 56. 1901, nom. nud. ORIGINAL MATERIAL: [Greenland, Ostgrönlands] Hurry-Inlet [...] [Leg. A.G. Nathorst, 1899].

Grimmia anodon var. *moravica* Podp., Věstn. Klubu Přír. v Prostějově 7: 18. 1905. PROTOLOGUE: [Czech Republic. Moravia, not explicitly stated].

Grimmia anodon var. *taeniophyllum* Mikut. ex Loeske, nom. inval., orthogr. err., Laubm. Eur. Part I: 45. 1913.

NOTE: Error in Wijk et al. (1969: 648) for *Schistidium angustum* var. *taeniophyllum* Mikut., Bryoth. Balt. 1909.

Grimmia apiculata var. *mutica* Schultz, Syll. Pl. Nov. 2: 135. 1828. PROTOLOGUE: [no locality given].

Grimmia apiculata var. *obtusa* Schimp., Coroll. Bryol. Eur.: 47. 1856. PROTOLOGUE: [no locality given].

Grimmia argyrotricha Müll. Hal., Flora 73(5): 485. 1890. PROTOLOGUE: [Tanzania] Africa or. trop., monte Kilima-Ndscharo, 3000–4800 m: Dr. Hans Meyer 1880.

Grimmia arvernica H. Philib., Rev. Bryol. 9: 25. 1882. *Grimmia plagiopodia* var. *arvernica* (H. Philib.) Boulay, Musc. France 1: 389. 1884. *Grimmia plagiopodia* f. *arvernica* (H. Philib.) Loeske, Biblioth. Bot. 101: 83. 1930. PROTOLOGUE: [France] [...] trouvée par M. Gautier Lacroze au Mont-Juzet, près de Clermont, à une altitude dènviron 430 mètres, sur un mur formé de blocs de lave et de mortier.

Grimmia austropulvinata Müll. Hal., Hedwigia 37: 160. 1898. PROTOLOGUE: Nova Seelandia, insula septentrionalis, prope Auckland ad rupes montis Wellington, Aug. 1882: T.F. Cheeseman.

Grimmia bifrons De Not., Comment. Soc. Crittog. Ital. 2: 102. 1866. *Grimmia alpestris* subsp. *bifrons* (De Not.) Podp., Consp. Musc. Eur.: 282. 1954, nom. inval. (ICBN Art. 34.2). *Grimmia donniana* var. *bifrons* (De Not.) Podp., Consp. Musc. Eur.: 280. 1954, nom. inval. (ICBN Art. 34.2). *Grimmia donniana* subsp. *bifrons* (De Not.) Wijk & Margad., Taxon 9: 190. 1960. PROTOLOGUE: [Italy] Dai monti della Sardegna centrale.

Grimmia caespiticia var. *bornmuellerorum* Schiffn., Oesterr. Bot. Z. 58: 312. 1908. PROTOLOGUE: [Iran. Tehran: Damavand] Persia bor.: Demawend, in Lavageröll unterhalb des Gipfels sehr verinzelt als pflanzeliche Spuren, 4000–5400 m s. m; 17.VII.1902 (Bornm., Iter Pers. alt. 1902, Nr. 5898).

Grimmia camonia Rota ex De Not., Atti Reale Univ. Genova 1: 755. 1869. PROTOLOGUE: [Italy] Crepacci delle rupi umide del monte Tonale.

Grimmia campestris f. *longirostrata* Vilh., Věstn. Král. Ceské Společn. Nauk, Tř. Mat.-Přír. 1924: 14. 1924. PROTOLOGUE: [Czech Republic] Habitat ad rupes arenosas in Bohemia prope Kralupy n. Vl., ad rupes quarzosas apud Štěchovice prope Pragam (Vel. [Velenovský]).

Grimmia campestris var. *lutescens* Mönk., Laubm. Eur.: 356. 1927. *Grimmia laevigata* var. *lutescens* (Mönk.) Podp., Consp. Musc. Eur.: 276. 1954. PROTOLOGUE: [Italy] Erhielt ich von Mauern in Bordighera in Norditalien (leg. Alwin Berger).

Grimmia campylotricha Müll. Hal., Flora 71: 414. 1888. PROTOLOGUE: Monte Kilimandscharo, ad finem sylvestrem superiorem inter 3000–4000 m.

Grimmia canescens Schleich. ex Spreng., Mant. Prim. Fl. Hal.: 55. 1807. *Grimmia ovata* var. *canescens* (Spreng.) Brid., Bryol. Univ. 1(1): 177. 1826. PROTOLOGUE: In rupibus Vallesiae. Schleicher.

Grimmia cardotii var. *gymnostoma* J. Froehl., Ann. Naturhist. Mus. Wien 67: 156. 1964. PROTOLOGUE: Afghanistan, Prov. Kabul, in valle Pagman prope Kabul, ca. 34°36'N, 68°56'E, ca. 2300–2800 m, substr. "Gneiß"; 21.VI.1962: nº 19.704.

Grimmia cardotii × *G. anodon* Culm., Rev. Bryol. 53: 8. 1926, nom. inval., published as a formula.

Grimmia cernua Nees & Hornsch., Bryol. Germ. 2(1): 174, tab. XXIV fig. 23a. 1827. PROTOLOGUE: Wächst auf dem Rauriser-Tauern am sogenannten Draba-Felsen in Gesellschaft mehrerer Gattungsverwandten, und wurde daselbst im Jahr 1816 zuerst entdeckt (Hoppe! Hsch.), in der Gastein (Mielichhofer! Funck!), beim Kalserthor in der Nahe des Glockners (Funck!), am ersten Leiterkopf auf der Pasterze, auf der Zirknitz-Alpe bei Dollach, dem Rathhausberg in der Gastein und dem Goldberg in der Raurifs (Laurer!), auf der Hohe der Scheidecker-Alpe im Oberhaslithal in der Schweiz (Hsch.).

Grimmia coarctata Müll. Hal., Hedwigia 37: 159. 1898. PROTOLOGUE: Nova Seelandia, insula australis, Canterbury, Lyttleton, ad rupes: T.W. Naylor Beckett, 1892; T.F. Cheeseman 1889 in Hb. Levier.

Grimmia commutata f. *basaltica* Vilh., Věstn. Král. Ceské Společn. Nauk, Tř. Mat.-Přír. 1924: 15. 1924. PROTOLOGUE: [Czech Republic] Habitat in Bohemia ad rupes basalticas (Vinařická hora, 1897, Vel.[Velenovský]).

Grimmia commutata f. *subepilosa* Braun-Blanq., Rev. Bryol. Lichénol. 23: 102. 1954, nom. nud. ORIGINAL MATERIAL: [Morocco] Ourgouz, 2.450 m., Si Zerekten [...]; Ourgouz, 2.500 m.

Grimmia commutata var. *brevipila* Warnst., Krypt.-Fl. Brandenburg 1(2): 294. 1904. *Grimmia commutata* f. *brevipila* (Warnst.) Vilh., Věstn. Král. Ceské Společn. Nauk, Tř. Mat.-Přír. 1924: 16. 1924. PROTOLOGUE: [Switzerland] sammelte Prof. Ascherson bei Locarno (Schweiz) an trockenen Gneissfelsen am Wege nach Ponte Brolla 2. Okt. 1885.

Grimmia commutata var. *gibba* Boulay, Musc. France 1: 367. 1884. *Grimmia ovalis* var. *gibba* (Boulay) Podp., Consp. Musc. Eur.: 279. 1954. PROTOLOGUE: [France] ça et là, Vosgues, St. Etienne (Pierrat); Pelvoux, Grandvillard près Briançon (B.); Mt-Louis (Renauld).

Grimmia commutata var. *maroccana* Meyl., Bull. Soc. Sci. Nat. Maroc 11: 115. 1931, nom. nud.

Grimmia commutata var. *microcarpa* Lesq., Mém. Soc. Sci. Nat. Neuchâtel 3(3): 27. 1846, nom. nud. ORIGINAL MATERIAL: sur les blocs erratiques de Rio, Val-de-Travers. Mai [L. Lesquereux].

Grimmia commutata f. *minor* Laz., Ukrajins'k. Bot. Zhurn. 3: 12. 1926. PROTOLOGUE: in Ukraïnia in districto Mariupol, Kamjani Mohyly in rupibus graniticis. 16.IX.1925 [leg. A.S. Lazarenko].

Grimmia commutata f. *rivularis* Loeske, Moosfl. Harz.: 186. 1903, nom. nud. *Grimmia commutata* f. *rivularis* Loeske, Laubm. Eur. Part I: 87. 1913. *Grimmia ovalis* f. *rivularis* (Loeske) Podp., Consp. Musc. Eur.: 279. 1954. PROTOLOGUE: [Germany] [...] im unteren Bodetal des Harzes [...] [Leg. L. Loeske].

Grimmia compacta Kindb., in Macoun, Cat. Canad. Pl. 7: 222. 1902, nom. nud.

Grimmia compactula Müll. Hal., Hedwigia 37: 167. 1898. PROTOLOGUE: Australia, Victoria, Grampians, Mt. Kosciusco: Dr. Sullivan 1884 in Hb. Melbourne.

Grimmia crassifolia var. *cucullata* Baumgartner, Ann. Naturhist. Mus. Wien 50: 534. 1939. PROTOLOGUE: Mittleres Elburs-Gebirge: Berge um Kalak nächst Keredj, ca. 1400 m (2283). TYPE: [Iran] Persia: Montes Elburs [Elburz] centr., in ditione oppidi Keredj, in montibus ad pagum Kalak. 20.V.1937 leg. K.H. Rechinger fil. nº 2283. Det. J. Baumgartner. (isotype, H!).

NOTE: The isotype in H is sterile, and in this condition is not possible to separate *G. tergestina* from *G. poecilostoma*.

Grimmia crinita var. *brevis* Boulay, Musc. France 1: 388. 1884. *Grimmia crinita* f. *brevis* (Boulay) Podp., Consp. Musc. Eur.: 275. 1954. PROTOLOGUE: [France, not explicitly stated].

Grimmia crinita var. *calabra* Zodda, Malpighia 23: 48. 1909. PROTOLOGUE: [Italy] Raccolta con frutti in aprile 1908 sulle rupi arenoso-calcaree sopra Melito Porto Salvo, 150 m., dal Prof. Nicotra.

Grimmia crinita f. *compacta* Besch., Cat. Mouss. Alg.: 16. 1882. PROTOLOGUE: [Algeria] Mont de Sel, Djebel Gharribou, El-Outaïa (La Perr.[De La Perraudière], mars 1858).

Grimmia crinita var. *elongata* Bruch & Schimp., Bryol. Europ. (fasc. 25 28) 3: 112, tab. 237ß. 1845. *Grimmia perichaetialis* Durieu & Coss., in Besch., Cat. Mouss. Alg.: 17. 1882, nom. inval., pro synon. PROTOLOGUE: [France] ad muros calcareos Monspeliae.

Grimmia crinita var. *laevis* Boulay ex Pit., Mém. Soc. Bot. France 54: 13. 1907, nom. nud. ORIGINAL MATERIAL : [Spain] Ténérife: Cañadas du pie de Teyde, entre El Portillo et Montaña Blanca, 2000 à 2500 m.

Grimmia curvifolia Lindb. ex Kindb., Bot. Not. 1882: 146. 1882, nom. nud.

Grimmia decipiens f. *falcata* Luisier, Mem. Real Acad. Ci. Exact. Madrid, ser. 2, 3: 111. 1924. PROTOLOGUE: [Spain. Salamanca] Sq.: [Sequeros] Las Batuecas.

Grimmia decipiens var. *homomalla* Bryhn, Nyt Mag. Naturvidensk. 32: 120. 1892. *Grimmia schultzii* var. *homomalla* (Bryhn) Paris, Index Bryol. ed. 2: 287. 1904. *Grimmia decipiens* f. *homomalla* (Bryhn) I. Hagen, Kongel. Norske Vidensk. Selsk. Skr. (Trondheim) 1909(5): 43. 1909. PROTOLOGUE: [Norway] Insula Tjoemo in ditione oppidi Toensberg ad rupes gneissicas interdum irroratas prope Kjaerelia socia *Jungermanniae autumnalis* var *subapicali* (4 Aprilii 1886).

Grimmia decipiens f. *major* Luisier, Mem. Real Acad. Ci. Exact. Madrid, ser. 2, 3: 111. 1924. PROTOLOGUE: [Spain] S.: Salamanca: Montalbo.

Grimmia donniana var. *curviseta* Lesq. & James, Man.: 143. 1884. PROTOLOGUE: Monitor, California (Lapham); Yakima River, Cascade Mountains (Watson).

Grimmia donniana f. *longiurna* Vilh., Věstn. Král. Ceské Společn. Nauk, Tř. Mat.-Přír. 1924: 18. 1924. PROTOLOGUE: [Czech Republic] Habitat in montibus Corconticis et Tatris. [...] Krkonoších (vrchol Sněžky, Labský vodopád, Obří důl, 1919, Vilh.) a ve Vys. Tatrách (u hotelu Kolbach, 1919, Vilh.).

Grimmia donniana f. *sudetica* Chal., Pamietn. Fizyogr. 2: 56. 1882. *Grimmia donniana* f. *breviseta* Vilh., Věstn. Král. Ceské Společn. Nauk, Tř. Mat.-Přír. 1924: 18. 1924, nom. illeg., incl. typ. f. prior. PROTOLOGUE: [Poland] Siodełko między Bystrą a Kamienistą od zachodu, na skałach pokruszonych; bardzo obficie. A także grań od siodełka aż pod sam szczyt Kamienistéj. Rohacze Zachodnie i Średnie; granie i pod same szczyty, w wielu miejscach; Pod samą czubą Wołowca i granie ku Klinowi. Kondracka granie ku Czerwonym Wirchom. Końcysta, od strony Tupéj.

Grimmia drakenbergensis Sim, Trans. Roy. Soc. South Africa 15: 209. 1926. PROTOLOGUE: [South Africa] Natal. Top of Giant's Castle, R.E. Symons (Sim 9962); Mont aux Sources, Sim 9963, and Miss Edwards.

Grimmia dusenii Paris, Index Bryol. 2: 524. 1895, nom. nud. *Guembelia dusenii* Müll. Hal. ex Paris, Index Bryol. 2: 524. 1895, nom. nud., pro synon. ORIGINAL MATERIAL: *Gümbelia Duseni* C.M., in Dusen M. Camer. nº 320. - Afr.: M. Cameroon.

Grimmia duthieana Müll. Hal. ex Paris, Index Bryol. Suppl.: 173. 1900, nom. nud.

Grimmia edinensis Fergusson ex W. Bell & J. Sadler, J. Bot. 8: 301. 1870, nom. inval., not accepted by authors. *Grimmia subsquarrosa* var. *edinensis* Braithw., Brit. Moss Fl. 2: 22. 1888. *Grimmia trichophylla* var. *edinensis* (Braithw.) Dixon, Stud. Handb. Brit. Mosses ed. 3: 148. 1924. PROTOLOGUE: [United Kingdom] Arthur's seat (Rev. J. Fergusson 1869).

Grimmia elatior f. *asperula* Loeske, Biblioth. Bot. 101: 190. 1930, nom. nud.

Grimmia elatior var. *physocarpa* De Not., Atti Reale Univ. Genova 1: 686. 1869. PROTOLOGUE: [Italy] circa Riva ad Sessitem[,] Carestia, in monte Musiné prope Taurinum[,] DNtrs.

Grimmia elatior var. *pseudofunalis* Limpr., Flora 65: 204. 1882. *Grimmia elatior* var. *subfunalis* Kindb., Bih. Kongl. Svenska Vetensk.-Akad. Handl. 7(9): 110. 1883, nom. inval., pro synon. PROTOLOGUE: [...] aus den Alpen [...], meist sterile Form Schlesiens und des nördlichen Europas [...].

Grimmia elatior f. *subinermis-asperula* Geh., Flora 62: 475. 1879. *Racomitrium asperulum* Geh., Flora 62: 475. 1879, nom. inval. *Grimmia elatior* var. *asperula* Limpr., Laubm. Deutschl. 1: 772. 1889, nom. illeg. *Grimmia asperula* Paris, Index Bryol. 2: 519. 1895, nom. inval., pro synon. *Grimmia elatior* var. *alpigena* H. Möller, Ark. Bot. 26A(2): 111. 1934, nom. illeg., incl. typ. var. prior. PROTOLOGUE: [Russia] Burgasutai-Pass, in einem kleinen Rasen, mit 7, 8 und 10 [*Barbula asperifolia*, *Barbula ruralis* and *Hedwigia ciliata*, respectively]; Kunowalski-Jurti, sehr spärlich zwischen *Hypnum vaucheri*. - Steril.

Grimmia elatior f. *sublaevis* Latzel, Hedwigia 66: 140. 1926. PROTOLOGUE: [Austria] V [Villach]: Görlitzen 800 m.

Grimmia elatior var. *submutica* Holler, Ber. Naturwiss. Vereins Schwaben Neuburg 30: 237. 1893. *Grimmia elatior* f. *submutica* (Holler) Podp., Consp. Musc. Eur.: 291. 1954. PROTOLOGUE: [Germany] Retterschwang-Thal bei der Alpe "in der Stellen" fruchtend 1364 m.

Grimmia elongata f. *epilosa* Stolz ex Matousch., Ber. Naturwiss.-Med. Vereins Innsbruck 28: 86. 1903. PROTOLOGUE: [Austria. Tirol] Möserlingwand bei Windisch-Matrei, 2000m, 13.IX.1896.

Grimmia elongata f. *epilosa* Limpr., Laubm. Deutschl. 1: 754. 1889, nom. nud. *Grimmia elongata* f. *epilosa* Vilh., Věstn. Král. Ceské Společn. Nauk, Tř. Mat.-Přír. 1924: 40. 1924, nom. illeg., non Matousch. 1903. PROTOLOGUE: Habitat in rupibus humidis rarissime in Alpibus et Tatris ([Slovakia] Malá studená dolina et apud 5 ples, 1919, Vilh.).

Grimmia elongata var. *patula* Bruch & Schimp., Bryol. Europ. (fasc. 25–28) 3: 130, tab. 259ß. 1845. *Grimmia patula* (Bruch & Schimp.) Schimp., Syn. Musc. Eur.: 223. 1860, comb. inval., pro synon. *Grimmia elongata* f. *patula* (Bruch & Schimp.) Vilh., Věstn. Král. Ceské Společn. Nauk, Tř. Mat.-Přír. 1924: 40. 1924. PROTOLOGUE: [Austria] locis siccis Messerlingswand Alpium tyrolensium (Hornsch.).

Grimmia elongata f. *propagulifera* Vilh., Věstn. Král. Ceské Společn. Nauk, Tř. Mat.-Přír. 1924: 40. 1924. PROTOLOGUE: Habitat in Tatris ([Slovakia] Mlinica supra lacum Skok, 1900 m, 1919, Vilh. [J. Vilhelm]).

Grimmia encalyptoides Müll. Hal. ex Watts & Whitel., Proc. Linn. Soc. New South Wales, ser. 2, 27: 83. 1902, nom. nud. ORIGINAL MATERIAL: [Australia. Victoria] Vic. Dr. Sullivan.

Grimmia filicaulis Müll. Hal., Nuovo Giorn. Bot. Ital., n.s. 3: 108. 1896. PROTOLOGUE: [China] In montibus Iu-kia-po, ad saxa, 24 Junio 1894.

Grimmia fuliginea Stirt., Ann. Scott. Nat. Hist. 17: 171. 1908. PROTOLOGUE: [United Kingdom] [...] near Arisaig.

Grimmia funalis f. *longipila* Boulay, Musc. France 1: 382. 1884. PROTOLOGUE: [not explicitly stated, localities for *G. funalis* s.l.] Pyrénées, [...], de la vallée d'Eynes (Arnott), jusqu'a Cauterets (Spruce), [...]; Plateau-Central, Mt.-Dore (Lamy); Alpes, Chanrousse, env. du lac Coeurzet (Ravaud), grande Chartreuse (de la Perraudière), C. sous les formes très diverses dans le massif du Mt-Blanc (Payot).

Grimmia funalis f. *pilosa* Loeske, Laubm. Eur. Part I: 155. 1913. PROTOLOGUE: [Austria] Klippen der Bachufer z.B. Stilluptal b. Mayrhofen, 1000 m.

Grimmia funalis f. *subepilosa* Mönk., Laubm. Eur.: 366. 1927. *Grimmia funalis* f. *subpilosa* Mönk. ex Podp., Consp. Musc. Eur.: 286. 1954, nom. inval., orthogr. err. PROTOLOGUE: [...] im Alpengebiete, in der Tatra, in den Karpathen, in Großbritannien und im nördlichen Europa. Im engeren Gebiete im Riesengebirge, im Arbergebiete des Bayerischen Waldes, im Algäu, Schwarzwalde, in de Vogesen und bei 400 m im Bodetale im Harze von Hampe beobachtet.

Grimmia funalis var. *cernua* Meyran, Ann. Soc. Bot. Lyon 39: 51. 1916, nom. nud. ORIGINAL MATERIAL: Savoie: Aiguilles de la Glière (Payot).

Grimmia funalis var. *epilosa* Milde, Bryol. Siles.: 148. 1869. *Grimmia funalis* f. *epilosa* (Milde) Mönk., Laubm. Eur.: 366. 1927. PROTOLOGUE: [Poland] Kesselgrube 3800' im Riesengebirge.

Grimmia funalis var. *laxa* Schimp., Syn. Musc. Eur.: 212. 1860. *Grimmia spiralis* var. *laxa* (Schimp.) Venturi, Rev. Bryol. 6(4): 55. 1879. PROTOLOGUE: in val di furva retro Bormio Vallis-tellinae.

Grimmia gebhardii Spreng., Mant. Prim. Fl. Hal.: 54. 1807. PROTOLOGUE: [Germany] In monte Baldo. Gebhard.

Grimmia hartmanii f. *epilosa* Milde, Bryol. Siles.: 150. 1869. *Dryptodon hartmanii* var. *epilosus* (Milde) Limpr., Laubm. Deutschl. 1: 791. 1889. PROTOLOGUE: [Poland, no locality explicitly stated for the variety] Königsberg, bei Kleinheide. (Sanio).– Schlesien: auf erratischen Blöcken der Ebene so bei Sagan (Everken); bei bunzlau am Teufelswehr bei Wehrau 500'; im Schusterbusche bei Klitschdorf und in den Looswitzer Steinbrüchen; bei Löwenberg am Plagwitzer Steinberge und auf dem Probsthainer Spitzberge (Limpr.); Jauergberg; Warthaberg und Volpersdorf (Ders.); Görbersdorf bei Waldenburg (Milde); Geiersberg und Zobten (Schulze); Ogulje; Schönau; Kitzelberg bei Kauffung und Mertenstein bei Jauer (Milde); Cudowa (Ders.); Fürstenstein; sehr gemein im Riesengebirge: Petersdorf bei Warmbrunn; Hartenberg bei Schreiberhau; Kochel- und Zackenfall-Gegend; Agnetendorf; Krummhübel; Brükenberg; am langen Seiffen. Sieghübel bei Wittichhaus im Isergebirge. (Limpr.).- Karlsbrunn und Kessel im Gesenke.

Grimmia hartmanii f. *longipila* E. Bauer, Repert. Spec. Nov. Regni Veg. 50: 342. 1941, nom. nud.

Grimmia hartmanii f. *robusta* E. Bauer, Repert. Nov. Spec. Regni Veg. 50: 342. 1941, nom. nud.

Grimmia hartmanii f. *secunda* Loeske, Biblioth. Bot. 101: 179. 1930. PROTOLOGUE: [Austria] [...] J. Murr in der Mellachschlucht (Vorarlberg, 620 m).

Grimmia hartmanii subsp. *bullata* Luisier, Brotéria Ci. Nat. 16: 90. 1947. PROTOLOGUE: [Portugal] Curral das Freitas: Pico Jorge, sur les rochers de basalte, 3 janv. 1945 (Nóbrega).

Grimmia hartmanii var. *crispa* Röll, Verh. Zool.-Bot. Ges. Wien 47: 662. 1897. *Dryptodon hartmanii* var. *crispus* (Röll) Dalla Torre & Sarnth., Fl. Tirol 5: 273. 1904. *Grimmia hartmanii* f. *crispa* (Röll) Podp., Consp. Musc. Eur.: 289. 1954. PROTOLOGUE: [Austria] Varenefall bei Riva am Gerdasee.

Grimmia hartmanii var. *fastigiata* Thér. & Mong., Bull. Soc. Agric. Sarthe 1899: 135. 1899. PROTOLOGUE: [France] St-Victeur: rochers siliceux à fleur de terre, à droite et à gauche de la route du Petit-Oisseau, en sortant du bourg. Abondante. Trouvée par M. E. Monguillon le 18 octobre 1893.

Grimmia hartmanii var. *japonica* Broth. ex Sasaoka, Bot. Mag. (Tokyo) 35(420): 272. 1921, nom. nud.

Grimmia hartmanii var. *mollis* Röll, Hedwigia 36: 38. 1897. *Dryptodon hartmanii* var. *mollis* (Röll) Podp., Consp. Musc. Eur.: 291. 1954, comb. inval., pro synon. PROTOLOGUE: [Spain] Sierra Guadarama [sic; Guadarrama], Pieta [sic; Siete] Picos [Dr. Dieck, nº] 240.

Grimmia hartmanii var. *montenegrina* Breidl. & Szyszył., Pl. Cernagor. Lect.: 29, tab. 2. 1888. *Dryptodon hartmanii* var. *montenegrinus* (Breidl. & Szyszył.) Limpr., Laubm. Deutschl. 1: 791. 1889. *Dryptodon simicii* Katić, in Loeske, Laubm. Eur. Part I: 139, fig. 43. 1913 ["simici"], nom. inval., pro synon. *Grimmia hartmanii* var. *meridionalis* Loeske, Laubm. Eur. Part I: 144. 1913, nom. illeg., incl. typ. var. prior. *Grimmia hartmanii* subsp. *montenegrina* (Breidl. & Szyszył.) Giacom., Atti Ist. Bot. Lab. Crittog. Univ. Pavia, ser. 5, 4: 223. 1947. PROTOLOGUE: Ad terram lapidosam silvae Perucica sub monte Kom Kucki ad confinium Albaniae borealis [leg. I. Szyszyłowicz].

Grimmia hawaiica Müll. Hal., Flora 82: 454. 1896. PROTOLOGUE: [U.S.A.] Insulae Hawaiicae, East Maui, in monte vulcanico Haleakala: Dr. Hillebrand 1872.

Grimmia hemipolia Stirt., Ann. Scott. Nat. Hist. 11: 109. 1902. PROTOLOGUE: [United Kingdom] [...] from Rannoch and Broadford in Skye [J. Stirton].

Grimmia heterophyllacea Müll. Hal. ex Paris, Index Bryol. Suppl.: 173. 1900, nom. nud.

Grimmia histrioipila Müll. Hal. ex Watts & Whitel., Proc. Linn. Soc. New South Wales, ser. 2, 27: 84. 1902, nom. nud. ORIGINAL MATERIAL: [Australia. Victoria] Vic., among lichens, Mt. Arapiles: Reader, Aug. '96.

Grimmia hornii Stirt., Scott. Naturalist (Perth) 10: 218. 1890. PROTOLOGUE: [United Kingdom] on bare rocks in Glen Ogle (G. Horn).

Grimmia hyalinocuspidata var. *mutica* Cardot & Broth., Kongl. Svenska Vetenskapsakad. Handl. 63(10): 26. 1923. PROTOLOGUE: [South Georgia] Georgia austr.: Bay of Isles, Rosita Harbour (38 p.p.).

Grimmia inaequalis Stirt., Ann. Scott. Nat. Hist. 17: 172. 1908. PROTOLOGUE: [United Kingdom] On stone walls, Garscube near Glasgow; near Forth Bridge at Dalmeny (J. M'Andrew).

Grimmia incana Wilson, Hooker's J. Bot. Kew Gard. Misc. 9: 323. 1857, nom. nud. ORIGINAL MATERIAL: Tibet occid. temp.; Rondu, alt. 6,000 ped. T.T. [Thomas Thompson nº 281].

Grimmia incurva f. *breviseta* Latzel, Hedwigia 66: 139. 1926. PROTOLOGUE: [Austria] L [Lavantquellengebiet]: Ameringkogel.

Grimmia incurva f. *longipila* Vilh., Věstn. Král. Ceské Společn. Nauk, Tř. Mat.-Přír. 1924: 32. 1924. PROTOLOGUE: [Slovakia] Habitat ad saxa in montibus Corconticis (Sněžka sub Obří bouda, 1450 m, 1919, Vilh. [J. Vilhelm]).

Grimmia incurva var. *pehrii* Latzel, Hedwigia 66: 140. 1926. PROTOLOGUE: [Austria] K [Koralpe]: Steinschober 1900 m (Gneiss).

Grimmia jacquinii Garov., Bryol. Austriac. Exc.: 25. 1840. PROTOLOGUE: [Imperio Austriaco].

Grimmia jaeggliana Giacom., Atti Ist. Bot. Lab. Crittog. Univ. Pavia, ser. 5, 9: 198. 1950. *Grimmia tergestina* var. *jaeggliana* (Giacom.) Podp., Consp. Musc. Eur.: 278. 1954. PROTOLOGUE: [Italy] Val di Susa (Piemonte) su rupi secche silicee presso Pradonio [...]; tra Bussoleno e Chianoc su rupi silicee terrose in un "Crossidietum squamigeri" [...]; [Switzerland] Canton Ticino: rupi calcaree scure cristalline, intercalate fra rocce silicee, soleggiatissime, preso Castione; ad Arogno su rupi calcaree seche [...].

Grimmia laevidens Broth., Acta Soc. Sci. Fenn. 19(12): 88. 1892. *Grimmia caucasica* Lindb. ex Broth., Acta Soc. Sci. Fenn. 19(12): 88. 1892, nom. inval., pro synon. *Grimmia levidens* Broth. ex Loeske, Laubm. Eur. Part I: 107. 1913, nom. inval., orthogr. err. PROTOLOGUE: [Georgia] Cauc. major occ.: Radscha, ad fontes fluminis Dschandschachi Tskhali, ad saxa silicea regionis alpinae superioris.

Grimmia leucophaea var. *elongata* Machado-Guim., Brotéria, Sér. Bot. 15: 9. 1917. *Grimmia campestris* var. *elongata* (Machado-Guim.) Cout., Musci Lusit.: 56. 1917. *Grimmia laevigata* var. *elongata* (Machado-Guim.) Machado-Guim., Cat. Descr. Briol. Port.: 40. 1918. PROTOLOGUE: [Portugal] Sanatório de Manteigas. Agosto de 1916 [A.L. Machado].

Grimmia leucophaea var. *epilosa* Warnst., Verh. Bot. Vereins Prov. Brandenburg 27: 47. 1885. PROTOLOGUE: [Germany] Nr. [Neuruppin]: Auf erratischen Blöcken sehr selten.

Grimmia leucophaea var. *funaliformis* Györffy & Péterfi, Mag. Bot. Lapok 3: 132. 1904. *Grimmia laevigata* f. *funaliformis* (Györffy & Péterfi) Podp., Consp. Musc. Eur.: 276. 1954. PROTOLOGUE: [Romania. Siebenbürgen] Im "Valea Saca" Thale.

Grimmia leucophaea var. *stenocarpa* G. Roth ex J.J. Amann, Fl. Mouss. Suisse 2: 130. 1918. PROTOLOGUE: Allemagne.

Grimmia leucophaea var. *subrotunda* Müll. Hal., Syn. Musc. Frond. 1: 794. 1849. PROTOLOGUE: [Australia] ad Swan River Novae Hollandiae: J. Drummond. Vere.

Grimmia limbatula Müll. Hal., Nuovo Giorn. Bot. Ital., n.s. 3: 108. 1896. PROTOLOGUE: [Japan] ex sylvis Yokohamae (Fr. Schaal 1879).

Grimmia minutula Müll. Hal. ex Kindb., Enum. Bryin. Exot., Suppl. 2: 107. 1893, nom. nud.

Grimmia mollis f. *propagulifera* J.J. Amann, Bull. Soc. Vaud. Sci. Nat. 53: 93. 1921. *Hydrogrimmia mollis* f. *propagulifera* (J.J. Amann) Podp., Consp. Musc. Eur.: 274. 1954. PROTOLOGUE: [Switzerland] Valais: Langefluh sur Saasfee, 2800 m (Amann) (BH. 87.6.30).

Grimmia mollis subsp. *submersa* Kindb., Eur. N. Amer. Bryin. 2: 218. 1898. PROTOLOGUE: Rocks in alp. Rivulets r. Eur. Norway!; Kaurin.

Grimmia mollis var. *aquatica* Schimp., Syn. Musc. Eur.: 222. 1860. *Grimmia mollis* f. *aquatica* (Schimp.) Broth., Laubm. Fennoskand.: 174. 1923. *Hydrogrimmia mollis* f. *aquatica* (Schimp.) Loeske, Biblioth. Bot. 101: 45. 1930. PROTOLOGUE: [...] summi Col di Stelvio [Italy] versus Orteler alt. 8000' legi aestate 1840, et in summae Alpis Albula Rhaetiae [Austria] parte granitica aest. 1845 et 1855, nec non in Sierra Nevada Hispan. [Spain] ad rivul. nival. in fauce abrupta Borreguil de s. Geronimo alt. 8500', in m. Dovre [Norway] ad lapides rivulor. Alp. Storhöe [...] (C. Hartman).

Grimmia muehlenbeckii f. *atrata* Péterfi, Mag. Bot. Lapok 2: 293. 1903. PROTOLOGUE: [Romania] Im Szurduk-Passe.

Grimmia muehlenbeckii f. *epilifera* Molendo, Ber. Naturhist. Vereins Passau 10: 120. 1875. *Grimmia trichophylla* f. *epilifera* (Molendo) Podp., Consp. Musc. Eur.: 288. 1954. PROTOLOGUE: [Germany] Livinallongo zwischen der Ornellabachmündung und Araba auf Melafyrtuff in enger Schlucht bei 4300' beobachtet: Mdo. [L. Molendo] (Oktober 1863).

Grimmia muehlenbeckii f. *umbrosa* J.E. Zetterst., Öfvers. Förh. Svenska Vetensk.-Akad. 34: 70. 1878, nom. nud. ORIGINAL MATERIAL: in latere boreali-occidentali montis Högkullen.

Grimmia muehlenbeckii var. *mutabilis* Sanio, in Holler, Ber. Naturwiss. Vereins Schwaben Neuburg 30: 237. 1893. PROTOLOGUE: [Germany] Berggündlethal unter der Pointalpe 1299 m.

Grimmia muehlenbeckii var. *subsquarrosa* J.J. Amann, Fl. Mouss. Suisse 1: 92. 1918. PROTOLOGUE: [not explicitly stated].

Grimmia nana Schleich., Cat. Pl. Helv. ed. 4: 40. 1821, nom. nud. ORIGINAL MATERIAL: [Helvetia].

Grimmia oblonga Wilson, Hooker's J. Bot. Kew Gard. Misc. 9: 323. 1857, nom. nud. ORIGINAL MATERIAL: Sikkim-Himalaya alpina; Jongri, alt. 12,000 ped., J.D.H. [J.D. Hooker nº 316].

Grimmia obtusolinealis Müll. Hal., Flora 73: 484. 1890. PROTOLOGUE: [Tanzania] Africa or. trop., monte Kilima-Ndscharo, 3000–4800 m: Dr. Hans Meyer 1889.

Grimmia oertzeniana Schultz, Syll. Pl. Nov. 2: 134. 1828. PROTOLOGUE: [Italy] Habitat en monte Rosa Italiae, solo micaceo. Ab Oertzen.

Grimmia orbicularis var. *longipila* Husn., Muscol. Gall.: 134. 1887. *Grimmia orbicularis* f. *longipila* (Husn.) Podp., Consp. Musc. Eur.: 285. 1954. PROTOLOGUE: [France, not explicitly stated].

Grimmia orbicularis var. *theriotii* Corb., Rev. Bryol. 17(2): 21. 1890. PROTOLOGUE: [France] Saint-Pavace, près Le Mans (Sartne): sur les murs du cimetière autour de l'église; février-mai 1889 (leg. Thériot).

Grimmia orbicularis × *G. pulvinata* Vilh., Preslia 2: 145. 1922, nom. inval., published as a formula.

Grimmia ovalis f. *umbrosa* Vaccari ex Tosco, Webbia 28: 288. 1973, nom. nud. ORIGINAL MATERIAL: [Italy] 3647 Capanna Gnifetti (Vacc.); 3400 Grivola (Vacc.); 3350 Cresta dell'Eco (Vacc.); 3259 Punta di Leppe (Vacc.).- (3130 Piz Gallegione (Pfeffer)).

Grimmia ovata f. *epilosa* Jur. ex Vent., Musc. Trentino: 42. 1899, nom. nud. *Grimmia ovalis* var. *epilosa* Jur. ex Warnst., Krypt.-Fl. Brandenburg 1(2): 296. 1904. PROTOLOGUE: [Germany] Brüsenwalde auf Feldsteinmauern.

Grimmia ovalis var. *heteracra* I. Hagen, Kongel. Norske Vidensk. Selsk. Skr. (Trondheim) 1909(5): 27. 1909. *Grimmia ovata* var. *heteracra* (I. Hagen) Loeske, Biblioth. Bot. 101: 134. 1930, nom. inval., pro synon. *Grimmia ovalis* f. *heteracra* (I. Hagen) C.E.O. Jensen, Skand. Bladmossfl.: 258. 1939. PROTOLOGUE: [Norway] ST. [Sor-Trondelag] Selbu, Rolset på en vandreblok: H.[Hagen].

Grimmia ovalis var. *simplex* Vondr., Bull. Soc. Amis Sci. Lett. Poznań, Sér. D, Sci. Biol. 6: 122, fig. 7. 1966. PROTOLOGUE: Iraq: Naprdan, dist. Rowanduz, 5.6.1961, leg. E. Hadač. Typus: Herb. Vondráček, num. Hč B. 201, isotypus in BUH asservatur.

Grimmia ovata f. *aquatica* Matousch., Ber. Naturwiss.-Med. Vereins Innsbruck 28: 86. 1903. *Grimmia ovalis* f. *aquatica* (Matousch.) Podp., Consp. Musc. Eur.: 279. 1954. PROTOLOGUE: [Austria.Tirol] Fruchtend im Zwieselbache (Ötztal), 23.IV.1899.

Grimmia ovata f. *umbrosa* Vacc., Bull. Soc. Fl. Valdôt. 9: 76. 1914, nom. nud.

Grimmia ovata var. *curviseta* I. Hagen, Tromsø Mus. Aarsh. 21–22(1): 63. 1899. *Grimmia ovalis* var. *curviseta* (I. Hagen) Podp., Consp. Musc. Eur.: 279. 1954. PROTOLOGUE: [Norway] in Kristians amt beobachtet; auch auf Dovrefjeld kommt sie vor.

Grimmia ovata var. *subturbinata* Brid., Bryol. Univ. 1(1): 177. 1826. *Grimmia ovalis* var. *subturbinata* (Brid.) L.I. Savicz & Smirnova, Novosti Sist. Nizsh. Rast. 5: 288. 1968. PROTOLOGUE: [Russia] In Kamtschatka habitat ubi Redowski legit.

Grimmia paramattensis Müll. Hal., Hedwigia 37: 161. 1898. PROTOLOGUE: Australia, New South Wales, Paramatta: Whitelegge Oct. 1884: Hb. Melbourne mis. 1885.

Grimmia patens f. *atrata* Röll ex Loeske, Biblioth. Bot. 101: 190. 1930. PROTOLOGUE: Norwegen: bei Bergen, am Buarbrae.

Grimmia pennsylvanica var. *bestii* Grout, Bryologist 7: 6. 1904. *Grimmia bestii* Grout, Bryologist 7: 6. 1904, nom. inval. prov. PROTOLOGUE: Newfane [1902–1903].

Grimmia polita Stirt., Ann. Scott. Nat. Hist. 16: 172. 1907. PROTOLOGUE: [United Kingdom] At Glenfinnan, a railway station half-way between Arisaig and Fort-William [J. Stirton].

Grimmia pulvinata f. *basaltica* Vilh., Věstn. Král. Ceské Společn. Nauk, Tř. Mat.-Přír. 1924: 28. 1924. PROTOLOGUE: Habitat ad rupes basalticas in monte Řip in Bohemia centrali (1919, Fr. A. Novák).

Grimmia pulvinata f. *brevicapsula* Boulay, Musc. France 1: 386. 1884. PROTOLOGUE: [France, not explicitly stated].

Grimmia pulvinata f. *brevipila* Boulay, Musc. France 1: 386. 1884. *Grimmia pulvinata* var. *brevipila* (Boulay) Dalla Torre & Sarnth., Fl. Tirol 5: 262. 1904. PROTOLOGUE: [France, not explicitly stated].

Grimmia pulvinata var. *epilosa* Schimp. ex Milde, Bryol. Siles.: 146. 1869. PROTOLOGUE: [Germany] Bärwalde in der Mark. (Ruthe).

Grimmia pulvinata lusus *holotricha* Györffy, Magyar Bot. Lapok 5: 343. 1906. PROTOLOGUE: [Hungary. Csongrad] [...] auf Steinmauern in Kis-Zombor.

Grimmia pulvinata f. *latifolia* Vilh., Věstn. Král. Ceské Společn. Nauk, Tř. Mat.-Přír. 1924: 28. 1924. PROTOLOGUE: [Czech Republic] Habitat ad rupes calcareas, aridas prope Radotín apud Pragam (1919, Vilh. [J. Vilhelm]).

Grimmia pulvinata var. *laxa* Bott., Atti Soc. Crittog. Ital. 3: 110. 1883. PROTOLOGUE: [Italy] [...] venne parimente raccolta da Arcangeli nel maggio presso la foce del fiume Angitola sulle rupi.

Grimmia pulvinata var. *laxa* Röll, Hedwigia 56: 170. 1915, nom. illeg., non Bott. 1883. *Grimmia pulvinata* f. *laxa* (Röll) Podp., Consp. Musc. Eur.: 285. 1954, nom. illeg., non Bott. 1883.

Grimmia pulvinata var. *longepilosa* Matousch., Verh. Zool.-Bot. Ges. Wien 51: 191. 1901. PROTOLOGUE: [Italy] Občina bei Triest (Sterneck). - D. Inseln Brazza: Zwischen Neresi und St. Pietro (Ginzberger, 1895).

Grimmia pulvinata var. *longicapsula* Besch., Bull. Soc. Bot. France 12: 136. 1865, nom. nud. *Grimmia pulvinata* var. *longicapsula* Debat, Fl. Musc.: 122. 1874. *Grimmia pulvinata* f. *longicapsula* (Debat) Boulay, Musc. France 1: 386. 1884. *Grimmia montana* var. *longicapsula* Besch., in Podp., Věstn. Klubu Přír. v Prostějově 7: 19. 1905 [NOTE: Error in Wijk et al. (1962: 393)]. PROTOLOGUE: [France] Hyères.

Grimmia pulvinata var. *longipila* Schimp., Syn. Musc. Eur. 1: 206. 1860. *Grimmia pulvinata* f. *longipila* (Schimp.) Boulay, Musc. France 1: 386. 1884. PROTOLOGUE: [not explicitly stated].

Grimmia pulvinata f. *minor* Boulay, Musc. France 1: 386. 1884. *Grimmia pulvinata* var. *minor* (Boulay) Dalla Torre & Sarnth., Fl. Tirol 5: 262. 1904. PROTOLOGUE: [France, not explicitly stated].

Grimmia pulvinata f. *nigricans* Warnst., Krypt.-Fl. Brandenburg 1(2): 291. 1904. PROTOLOGUE: [Germany] bei Neuruppin auf der Mauer des neuen Kirchhofs.

Grimmia pulvinata subsp. *philibertii* Kindb., Enum. N. Amer. Bryin. 2: 219. 1897. PROTOLOGUE: [U.S.A.] Amer. r. Idaho: Leiberg.

Grimmia pulvinata f. *robusta* Boulay, Musc. France 1: 386. 1884. *Grimmia pulvinata* var. *robusta* (Boulay) Dalla Torre & Sarnth., Fl. Tirol 5: 262. 1904. PROTOLOGUE: [France, not explicitly stated].

Grimmia pulvinata f. *serpentinosa* Vilh., Spisy Přír. Fak. Karlovy Univ.: 33. 1925. PROTOLOGUE: [Czech Republic] [...] in Bohemia occid. ad Mnichov prope Mariánské Lázně.

Grimmia pulvinata var. *viridis* Schimp., Syn. Musc. Eur.: 206. 1860. *Grimmia pulvinata* f. *viridis* (Schimp.) Loeske, Laubm. Eur. Part I: 163. 1913. PROTOLOGUE: [France] ad St. Léonard prope Limoges unde cl. Lamy misit.

Grimmia retracta Stirt., Scott. Naturalist (Perth) 8: 235. 1886. *Grimmia hartmanii* f. *retracta* (Stirt.) Loeske, Laubm. Eur. Part I: 141. 1913. *Grimmia hartmanii* subsp. *retracta* (Stirt.) Dixon, Stud. Handb. Brit. Mosses ed. 3: 151. 1924. *Grimmia hartmanii* var. *retracta* (Stirt.) Podp., Consp. Musc. Eur.: 290. 1954. *Grimmia hausmani* subsp. *retracta* (Stirt.) Podp., Consp. Musc. Eur.: 290. 1954, nom. inval. PROTOLOGUE: [United Kingdom] Found beside Loch Tay, July, 1866 [J. Stirton].

Grimmia retracta var. *submutica* Stirt., Ann. Scott. Nat. Hist. 6: 121. 1897. PROTOLOGUE: [United Kingdom] Achnacarron, Loch Awe, 1896.

Grimmia robusta Braithw., J. Bot. 10: 196, tab. 124 fig. 3. 1872, nom. illeg., non Nees & Hornsch. 1827. *Grimmia ambigua* Braithw., J. Bot. 10: 196. 1872, nom. inval., pro synon. *Grimmia decipiens* var. *robusta* Braithw., Brit. Moss Fl. 2: 23. 1888, nom. nov. *Grimmia decipiens* subsp. *robusta* (Braithw.) Dixon, Stud. Handb. Brit. Mosses: 141. 1896. *Grimmia trichophylla* var. *robusta* (Braithw.) A.J.E. Smith, J. Bryol. 9: 279. 1976. *Grimmia brittanica* A.J.E. Smith, J. Bryol. 17: 272. 1992. PROTOLOGUE: [United Kingdom. Scotland] Rocks in Alpine districts. Clova mountains, at 2000 feet (Mr. Fergusson [1868]). Fairhead, Ireland (Dr. Moore). Cardross and Bowling (Dr. Stirton [1864]). Ross-shire (Mr. Hunt).

Grimmia rotunda R. Br. bis, Trans. & Proc. New Zealand Inst. 27: 416, pl. 32 fig. 22. 1895. PROTOLOGUE: [New Zealand] Rocks, Lyttelton hills; 1880. Collected by R.B.

Grimmia rubescens Stirt., Scott. Bot. Rev. 1: 90. 1912. PROTOLOGUE: [United Kingdom. Western Highlands, Ross-shire, near Gairloch, leg. Stirton].

Grimmia sardoa f. *propagulifera* M. Fleisch., Malpighia 7: 329. 1893. PROTOLOGUE: [Italy] Sardegna: Monte Santo, Pula, leg. Moris.

Grimmia serrulata F. Weber & D. Mohr, Index Mus. Pl. Crypt.: 2. 1803, nom. nud.

Grimmia stenophylla Müll. Hal., Hedwigia 37: 158. 1898. PROTOLOGUE: [Australia] Tasmania, ad rupes on the Sugar loaf, Green Ponts, Majo 1892: W.A. Weymouth in Hb. Brotheri, qui misit 1893 sub *G. trichophylla* Grev.

Grimmia stevensii R. Br. bis, Trans. & Proc. New Zealand Inst. 27: 418, pl. 34 fig. 30. 1895. PROTOLOGUE: [New Zealand] Mount Torlesse; January, 1886. Collected by R.B.

Grimmia stirlingii Müll. Hal., Hedwigia 37: 163. 1898. *Schistidium stirlingii* (Müll. Hal.) Ochyra, Fragm. Florist. Geobot. 43: 107. 1998. PROTOLOGUE: Australia, Victoria, Omeo alt. 2500–3000 pedum, ad rupes J. Stirling 1883 in Hb. Melbourne.

Grimmia subaquila Stirt., Ann. Scott. Nat. Hist. 18: 243. 1909. PROTOLOGUE: [United Kingdom] At Loch-nan-Uamb, several miles to the east of Arisaig and near Prince Charlie's Cave; on stones covered with earth, by the Rev. J.A. M'Caskill of Onich, September 1908.

Grimmia subcallosa Müll. Hal., Hedwigia 37: 161. 1898. PROTOLOGUE: Australia, Victoria, Mt. Cole Pyrenaeorum: D. Sullivan 1883 in Hb. Melbourne.

Grimmia subelongata Göpp. & Menge, Ber. Bekanntm. Verh. Königl. Preuss. Akad. Wiss. Berlin 18: 459. 1853, nom. nud.

Grimmia subfusca Wilson, Hooker's J. Bot. Kew Gard. Misc. 9: 323. 1857. PROTOLOGUE: [India] Khasia temperata; Kollong, alt. 5,000 ped., J.D.H. [J.D. Hooker n° 296].

Grimmia subsulcata f. *dominiana* Vilh., Věstn. Král. Ceské Společn. Nauk, Tř. Mat.-Přír. 1924: 24. 1924. *Grimmia alpestris* f. *dominiana* (Vilh.) Podp., Consp. Musc. Eur.: 282. 1954. PROTOLOGUE: [Poland] Habitat ad rupes graniticas supra lacum "Skalnaté pleso" (2000 m) sub montem Lomnický štít in Tatris (leg. 1919, Domin).

Grimmia subtergestina Müll. Hal., Nuovo Giorn. Bot. Ital., n.s. 3: 109. 1896. PROTOLOGUE: [China] Prov. Kansu, ad flumen Bardun prope ostium fluminis Solomo, Majo 1886: G.N. Potanin legit sterilem.

Grimmia tergestina f. *epilosa* Limpr., Laubm. Deutschl. 1: 740. 1889. PROTOLOGUE: [Austria] [...] sammelte H. Gander die ♂ Pflanze an senkrechten Kalkwänden über Kalchstein 1750 m bei Innervillgraten in Tirol.

Grimmia tergestina var. *carpatica* Vanek ex Podp., Consp. Musc. Eur.: 277. 1954, nom. nud.

Grimmia tergestina subsp. *tergestinoides* Culm., Rev. Bryol. 38: 75. 1911, nom. alt. *Grimmia tergestina* var. *tergestinoides* Culm., Rev. Bryol. 38: 75. 1911, nom. alt. PROTOLOGUE: [Switzerland] La Chaux prope Sainte-Croix, montis Jura helvetici, c. fructu unico, altitudine 1080 m. legit Ch. Meylan. (Sur calcaire aquitanien).

Grimmia torquata var. *pilifera* Scheutz, Öfvers. Förh. Kongl. Svenska Vetensk.-Akad. 27: 94. 1870, nom. nud. ORIGINAL MATERIAL: [Sweden] Skurugata.

Grimmia torquata-andreaeoides Loeske, in J.J. Amann, Bull. Soc. Vaud. Sci. Nat. 54: 45. 1921, nom. inval., published as a formula.

Grimmia tortilis Muhl., Cat. Pl. Amer. Sept.: 98. 1813, nom. nud.

Grimmia trichophylla f. *epilosa* Warnst., Verh. Bot. Vereins Prov. Brandenburg 39: 31. 1897. *Grimmia trichophylla* var. *epilosa* (Warnst.) Warnst., Krypt.-Fl. Brandenburg 1(2): 299. 1904. PROTOLOGUE: [Germany. Mark Brandenburg] An erratischen Blöcken an der Strasse nach Gr. Schönebeck.

Grimmia trichophylla var. *gemmipara* Husn., Fl. Mousses Nord-Ouest: 93. 1873. PROTOLOGUE: [France] Ségrie-Fontaine [Orne].

Grimmia trichophylla var. *longipila* Warnst., Krypt.-Fl. Brandenburg 1(2): 299. 1904. *Grimmia trichophylla* f. *longipila* (Warnst.) Mönk., Laum. Eur.: 367. 1927. PROTOLOGUE: [Poland. Postdam] Ostprignitz: Mansfeld, erratischer Block auf der Heide (Jaap!).

Grimmia trichophylla var. *lusitanica* Schimp., Syn. Musc. Eur. ed. 2: 257. 1876. PROTOLOGUE: [Portugal] in rup. graniticis m. Foya ad Monchique Algarviae (C.H. Solms).

Grimmia trichophylla var. *mutica* Grav. ex Delogne, Ann. Soc. Belge Microscop. 9: 178. 1885. PROTOLOGUE: [Belgium] Profondeville (Grav.).

Grimmia trichophylla var. *nigra* R. Br. bis, Trans. & Proc. New Zealand Inst. 27: 415, pl. 31 fig. 19. 1895. PROTOLOGUE: [New Zealand] Rocks in the gorge of the River Hurunui; March, 1893. Collected by R.B.

Grimmia trichophylla var. *submutica* Boulay, Musc. France 1: 379. 1884. *Grimmia trichophylla* f. *submutica* (Boulay) Corb., Mém. Soc. Sci. Nat. Cherbourg 26: 256. 1889. PROTOLOGUE: [France] grès désagrégés, frais dans l'Estérel (B.).

Grimmia trichophylla f. *vulgaris* Mönk., Laubm. Eur.: 367. 1927. PROTOLOGUE: ...der Hügel- und Bergregion durch Europa [...] deutschen Mittelgebirgen [...] Norddeutschen Ebene.

Grimmia triformis var. *brevifolia* Chal. ex J.J. Amann, Bull. Soc. Vaud. Sci. Nat. 55: 131. 1923, nom. nud. ORIGINAL MATERIAL: [Switzerland] Bern: Finsteraarhorngipfel, 4275 m. Frey.

Grimmia truncatoapocarpa Müll. Hal., Hedwigia 37: 163. 1898. *Schistidium truncatoapocarpum* (Müll. Hal.) Ochyra, Fragm. Florist. Geobot. 43: 107. 1998. PROTOLOGUE: Australia subtropica, New South Wales, Paramatta: Whitelegge. Octobri 1884. Hb. Melbourne misit 1885.

Grimmia uncinata Kaulf., in Sturm, Deutschl. Fl. Abtheilung II, Cryptogamie 4(15): 13. 1816. PROTOLOGUE: [Austria] Auf der Seethaler Alpe in Stenermark fand ich dieses Moos fehr häufig im August mit reisen Fruchten in Steinrißen.

Grimmia undulata Stirt., Scott. Bot. Rev. 1: 91. 1912. PROTOLOGUE: [United Kingdom. Scotland: Rossshire] On the red rock in several places near Gairloch [J. Stirton].

Grimmia unicolor f. *piligera* Molendo, in Lorentz, Moosstud.: 94. 1864, nom. nud. *Grimmia unicolor* var. *piligera* (Molendo) Grav., Rev. Bryol. 10: 24. 1883, comb. inval.

Grimmia versabilis R. Br. bis, Trans. & Proc. New Zealand Inst. 27: 415, pl. 32 fig. 20. 1895. PROTOLOGUE: [New Zealand] summit of Port Lyttelton hills; 1880. Collected by R.B.

Grimmia wrightii R. Br. bis, Trans. & Proc. New Zealand Inst. 27: 413, pl. 30 fig. 12. 1895, nom. illeg., non Austin 1875. *Schistidium wrightii* Ochyra, Fragm. Florist. Geobot. 43: 107. 1998 nom. nov. PROTOLOGUE: Limestone rocks, Broken River, Canterbury; November, 1889. Collected by R.B.

Grimmia zahlbruckneri Garov., Bryol. Austriac. Exc.: 26. 1840. PROTOLOGUE: [Imperio Austriaco].

Guembelia arctica Müll. Hal., Zweite Deut. Nordpolarf. 2(1): 73. 1873. *Grimmia arctica* (Müll. Hal.) A. Jaeger, Ber. Tätigk. St. Gallischen Naturwiss. Ges. 1872–73: 71 (Gen. Sp. Musc. 1: 353). 1874. *Racomitrium hypnoides* var. *arcticum* (Müll. Hal.) C.E.O. Jensen, Meddel. Grønland 3: 345. 1887. *Grimmia caespiticia* subsp. *arctica* (Müll. Hal.) Kindb., Eur. N. Amer. Bryin. 2: 220. 1898. PROTOLOGUE: [Greenland] Kaiser-Franz-Josephs-Fjord inter 73–74° lat. bor., unde cespes singularis relatus est.

Guembelia caucasica Müll. Hal., Bot. Zeitung (Berlin) 14: 421. 1856. *Grimmia caucasica* (Müll. Hal.) A. Jaeger, Ber. Tätigk. St. Gallischen Naturwiss. Ges. 1872-73: 72 (Gen. Sp. Musc. 1: 354). 1874. PROTOLOGUE: Caucasus: Relanger sub No. 9. Hb. Kunzeano communicavit.

Guembelia erythraea Müll. Hal., in Brizi, Malpighia 7: 296. 1893. *Grimmia erythraea* (Müll. Hal.) Cufod. Oesterr. Bot. Z. 98: 227. 1951. PROTOLOGUE: [Ethiopia] Abyssinia, prope Keren.

Racomitrium patens subsp. *subsimplex* Kindb., Boll. Soc. Bot. Ital. 1896(1): 22. 1896. *Grimmia patens* subsp. *subsimplex* (Kindb.) Kindb., Eur. N. Amer. Bryin. 2: 225. 1898. *Grimmia patens* subsp. *simplex* Kindb. ex Paris, Index Bryol. Suppl.: 294. 1900, nom. inval., orthogr. err. PROTOLOGUE: [Switzerland] Canton d'Uri [...] Goeschenen [N.C. Kindberg & J. Röll, 1895].

Racomitrium patens var. *crassifolium* Torka, Allg. Bot. Z. Syst. 9(7/8): 146. 1903. *Grimmia patens* f. *crassifolia* (Torka) Podp., Consp. Musc. Eur.: 292. 1954. PROTOLOGUE: [Germany] [...] am 27.August 1902 auf einem Steine bei Schwiebus [Saxonia] gefunden worden ist.

Schistidium teretinerve f. *lanceolatum* Podp., Veröff. Geobot. Inst. Rübel Zürich 1: 245. 1924. *Grimmia teretinervis* f. *lanceolata* (Podp.) Podp., Consp. Musc. Eur.: 281. 1954. PROTOLOGUE: [Switzerland] [...] Rande der Buffalorawiese.

Trichostomum pulvinatum var. *elongatum* Schultz, Syll. Pl. Nov. 2: 142. 1828. *Grimmia pulvinata* var. *elongata* (Schultz) Fiedl., Syn. Laubm. Mecklenb.: 82. 1844. PROTOLOGUE: [Germany] Habitat in saxis granitosis Fl. Stargard.

LITERATURE CITED

Abramova, A. L. & I. I. Abramov. 1988. De speciebus generis *Grimmia* Hedw. ex Asia media. Novosti Sist. Nizsh. Rast. 25: 157–169.

Allen, B. 1995. Eight neglected species of Grimmiaceae (Musci) from North America. Fragm. Florist. Geobot. 40: 159–166.

Bednarek-Ochyra, H., J. Muñoz & R. Ochyra. 1992. The identity of *Grimmia pyrenaica* (Musci, Grimmiaceae). Fragm. Florist. Geobot. 37: 389–393.

Blom, H. H. 1996. A revision of the *Schistidium apocarpum* complex in Norway and Sweden. Bryophyt. Biblioth. 49: 1–333.

Bremer, B. 1980 [1981]. A taxonomic revision of *Schistidium* (Grimmiaceae, Bryophyta) 2. Lindbergia 6: 89–117.

Bridel, S. E. 1806. Muscologia Recentiorum Supplementum 1: [i]–viii, [1]–271. Gothae. Carolum Guil. Ettingerum.

Brown, R. 1903. On the musci of the calcareous districts of New Zealand, with descriptions of new species. Trans. & Proc. New Zealand Inst. 35: 323–342, pl. 35–40.

Cao, T. & S. P. Churchill. 1995. New synonyms in *Grimmia* and *Schistidium* (Bryopsida: Grimmiaceae). Nova Hedwigia 60: 505–513.

Cao, T. & D. H. Vitt. 1986. A taxonomic revision and phylogenetic analysis of *Grimmia* and *Schistidium* (Bryopsida; Grimmiaceae) in China. J. Hattori Bot. Lab. 61: 123–247.

Churchill, S. P. 1981. A phylogenetic analysis, classification and synopsis of the genera of the Grimmiaceae (Musci). Advances Cladist. 1: 127–144.

Churchill, S. P. 1989. Bryologia novo granatensis. Estudio de los musgos de Colombia IV. Catálogo nuevo de los musgos de Colombia. Trop. Bryol. 1: 95–132.

Churchill, S. P. 1994. New synonyms and deletions for the moss floras of Colombia and Ecuador. Trop. Bryol. 9: 1–4

Crosby, M. R. & R. E. Magill. 1994. Index of Mosses 1990–1992. Monogr. Syst. Bot. Missouri Bot. Gard. 50: [i–vi], 1–87.

Crosby, M. R. & R. E. Magill. 1997. Index of Mosses 1993–1995. Monogr. Syst. Bot. Missouri Bot. Gard. 62: [i–vi], 1–106.

Crosby, M. R., R. E. Magill & C. R. Bauer. 1992. Index of Mosses 1963–1989. Monogr. Syst. Bot. Missouri Bot. Gard. 42: [i–vi], 1–646.

Crum, H. 1994. Grimmiales. *In* A. J. Sharp, H. Crum & P. M. Eckel (editors), The Moss Flora of Mexico. Part One. Sphagnales to Bryales. Mem. New York Bot. Gard. 69: 386–415.

Deguchi, H. 1978 [1979]. A revision of the genera *Grimmia*, *Schistidium* and *Coscinodon* (Musci) of Japan. J. Sci. Hiroshima Univ., Ser. B, Div. 2, Bot. 16: 121–256.

Deguchi, H. 1984. Studies on some Patagonian species of Grimmiaceae (Musci, Bryophyta). *In* H. Inoue (ed.), Studies on Cryptogams in Southern Chile: 17–72. Kensei-sha, Tokyo.

Deguchi, H. 1986. Notes on some Asian species of *Grimmia*. Hikobia 9: 327–329.

Deguchi, H. 1987. Studies on some Peruvian species of the Grimmiaceae (Musci, Bryophyta). *In* H. Inoue (ed.), Studies on Cryptogams in Southern Peru: 19–74. Tokai University Press, Tokyo.

Dixon, H. N. 1926. Studies in the bryology of New Zealand, with special reference to the herbarium of Robert Brown. Part IV. New Zealand Inst. Bull. 3: 153–161, 283, pl. 9.

Frisvoll, A. A. 1978. Nomenclatural note on *Trichostomum cuspidatissimum*. Bryologist 81: 156–159.

Geissler, P. & E. Maier. 1995. Lectotypifications of central European *Grimmia* species (Musci, Grimmiaceae). Candollea 50: 495–514.

Greven, H. C. 1997. *Grimmia austro-funalis* C. Müll., a species with a misleading name and a disjunct distribution. J. Bryol. 19: 827–830.

Greven, H. C. 1998. Synopsis of *Grimmia* Hedw. in New Zealand, including *Grimmia wilsonii sp. nov.* J. Bryol. 20: 389–402.

Ignatov, M. S. & O. M. Afonina. 1992. Check-list of mosses of the former USSR. Arctoa 1: 1–85.

Ignatov, M. S. & T. Cao. 1994. Bryophytes of the Altai Mountains. IV. The family Grimmiaceae. Arctoa 3: 67–122.

Jones, G. N. 1933. Grimmiaceae. *In* A. J. Grout (ed.), Moss Flora of North America 2: 1–66, pl. 1–25. Newfane, Vermont.

Magill, R. E. 1981. Bryophyta. Part I Mosses. Fascicle 1, Sphagnaceae–Grimmiaceae. *In* O. A. Leistner (ed.), Flora of Southern Africa. Department of Agriculture and Fisheries, Pretoria.

Maier, E. 1998. Zur systematischen Stellung von *Grimmia pitardii* Corb. (Musci). Candollea 53: 301–308.

Mårtensson, O. 1956. Bryophytes of the Torneträsk Area, Northern Swedish Lappland. II. Musci. Kung. Svenska Vetenskapsakad. Avh. Naturskyddsärenden 14: 1–321.

Matteri, C. M. & J. Muñoz. 2000. On the identity of *Grimmia depressa* and *G. depressa* var. *terrestris*. J. Bryol. 22: 62–63.

McIntosh, T. T. & J. R. Spence. 1986. *Grimmia olympica* Britt. (Grimmiaceae) is transferred to *Brachydontium* (Seligeraceae). Bryologist 89: 200–202.

Muñoz, J. 1997. The correct name of *Grimmia alpestris* (Musci, Grimmiaceae). Bryologist 100: 517–519.

Muñoz, J. 1998a. Materials toward a revision of *Grimmia* (Musci: Grimmiaceae): Nomenclature and taxonomy of *Grimmia longirostris*. Ann. Missouri Bot. Gard. 85: 352–363.

Muñoz, J. 1998b. A taxonomic revision of *Grimmia* subgenus *Orthogrimmia* (Musci, Grimmiaceae). Ann. Missouri Bot. Gard. 85: 367–403.

Muñoz, J. 1999a. A revision of *Grimmia* (Musci, Grimmiaceae) in the Americas. 1: Latin America. Ann. Missouri Bot. Gard. 86: 118–191.

Muñoz, J. 1999b. *Grimmia arcuatifolia* and *G. leibergii* (Musci, Grimmiaceae), two neglected species from northwestern North America. Anales J. Bot. Madrid 57: 7–13.

Muñoz, J. 2000a. (1450) Proposal to conserve the name *Grimmia poecilostoma* Cardot & Sebille (Musci, Grimmiaceae). Taxon 49: 287–288.

Muñoz, J. 2000b. (1451) Proposal to conserve the name *Grimmia crinita* (Musci, Grimmiaceae) against *Dicranum phascoideum*. Taxon 49: 289–290.

Muñoz, J. 2000c. New synonyms in *Grimmia* (Grimmiaceae). J. Bryol. 22: 99–102.

Muñoz, J. & R. Ochyra. 1999. *Grimmia australis*, new combination. J. Bryol. 21: 335–336.

Nyholm, E. 1998. Illustrated Flora of Nordic Mosses. Fasc. 4. Nordic Bryological Society, Copenhagen & Lund.

Ochyra, R. 1989a. Lectotypification of *Schistidium pulvinatum* (Hedw.) Brid. (Musci: Grimmiaceae) and its consequences. Nova Hedwigia 48: 85–106.

Ochyra, R. 1989b. Proposal to conserve *Schistidium* (Musci, Grimmiaceae) with a new type. Taxon 38: 665–669.

Ochyra, R. 1990. The discovery of the South Georgian endemic species *Schistidium urnulaceum* (C. Muell.) B.G. Bell (Musci: Grimmiaceae) in the Antarctic. Polish Polar Res. 11: 133–146

Ochyra, R. 1993. *Grimmia plagiopodia* (Musci, Grimmiaceae) in the Southern Hemisphere. Fragm. Florist. Geobot. 38: 21–27.

Ochyra, R. & H. Bednarek-Ochyra. 1994. A reconsideration of *Grimmia brandegei* Musci, Grimmiaceae. Fragm. Florist. Geobot. 392: 667–670.

Ochyra, R. & B. G. Bell. 1984. A record of *Schistidium falcatum* (Bryophyta: Musci) from the Antarctic. Bull. Brit. Antarc. Surv. 64: 77–79.

Ochyra, R. & G. Cisło. 1999. The moss herbarium of Titus Chałubiński in the Tatra Museum in Zakopane. Polish Bot. Stud., Guidebook Ser. 22: 1–178.

Ochyra, R. & H. Hertel. 1990. Contribution to the moss flora of the subantarctic Island Marion. Polish Bot. Stud. 1: 19–34.

Ochyra, R. & C. M. Matteri. 1996. Nomenclatural changes in South American and sub-Antarctic mosses. Fragm. Florist. Geobot. 41: 1005–1010.

Ochyra, R., C. Sérgio & R. Schumacker. 1988. *Racomitrium lamprocarpum* (C. Muell.) Jaeg., an austral moss disjunct in Portugal, with taxonomic and phytogeographical notes. Bull. Jard. Bot. État 58: 225–258.

Pfeiffer, L. 1873. Nomenclator botanicus 1(1): [i–iv], 1–808. T. Fisher, Cassellis [Kassel].

Pfeiffer, L. 1874. Nomenclator botanicus 1(2): [i], 809–1876. T. Fisher, Cassellis [Kassel].

Robinson, H. & F. J. Hermann. 1964. Notes on American *Grimmias*. Bryologist 67: 170–174.

Savicz-Lyubitzkaya, L. I. & Z. N. Smirnova. 1970. Opredelitel' Listostebel'nykh mkhov SSSR. Verkhoplodnye mkhi [The Handbook of the Mosses of the U.S.S.R. The Mosses Acrocarpous]. Nauka, Leningrad.

Takaki, N., T. Amakawa, T. Osada & E. Sakuma. 1970. Bryophyte flora of Mt. Kaikoma, Mt. Senjo and Mt. Mitadake. J. Hattori Bot. Lab. 33: 171–202.

Vaarama, A. 1961. The taxonomic status of *Grimmia flettii*. Bryologist 64: 320–325.

Wijk, R. van der, W. D. Margadant & P. A. Florschütz. 1962. Index Muscorum 2. D–Hypno. Regnum Veg. 26: 1–535.

Wijk, R. van der, W. D. Margadant & P. A. Florschütz. 1969. Index Muscorum 5. T–Z, Appendix. Regnum Veg. 65: [i]–xii, 1–922.

INDEX OF SCIENTIFIC NAMES

Accepted taxa are in **bold**. An asterisk (*) indicates lectotypification designated in this paper.